Manager's Guide
—————— to ——————
C E N T R E X

The Artech House Telecommunication Library

The Executive Guide to Video Teleconferencing by Ronald J. Bohm and Lee B. Templeton

The Telecommunications Deregulation Sourcebook, Stuart N. Brotman, ed.

Digital Cellular Radio   by George Calhoun

E-Mail Stephen A. Caswell

The ITU in a Changing World by George A. Codding, Jr. and Anthony M. Rutkowski

Design and Prospects for the ISDN by G. DICENET

Introduction to Satellite Communication by Bruce R. Elbert

Television Programming across National Boundaries: The EBU and OIRT Experience by Ernest Eugster

The Competition for Markets in International Telecommunications by Ronald S. Eward

A Bibliography of Telecommunications and Socio-Economic Development by Heather E. Hudson

New Directions in Satellite Communications: Challenges for North and South, Heather E. Hudson, ed.

Communication Satellites in the Geostationary Orbit by Donald M. Jansky and Michel C. Jeruchim

World Atlas of Satellites, Donald M. Jansky, ed.

Handbook of Satellite Telecommunications and Broadcasting, L. Ya. Kantor, ed.

World-Traded Services: The Challenge for the Eighties by Raymond J. Krommenacker

Telecommunications: An Interdisciplinary Text, Leonard Lewin, ed.
Telecommunications in the U.S.: Trends and Policies, Leonard Lewin, ed.
Introduction to Telecommunication Electronics by A.Michael Noll
Introduction to Telephones and Telephone Systems by A. Michael Noll
Teleconferencing Technology and Applications by Christine H. Olgren and Lorne A. Parker

The ISDN Workshop: INTUG Proceedings, G. Russell Pipe, ed.
Integrated Services Digital Networks by Anthony M. Rutkowski

The Law and Regulation of International Space Communication by Harold M. White, Jr. and Rita Lauria White

Manager's Guide
—— to ——
C E N T R E X

John R. Abrahams

Artech House

Library of Congress Cataloging-in-Publication Data

Abrahams, John B.
The manager's guide to Centrex.

1. Centrex telephone service. I. Title.
TK6430.A27   1988   384.6'42   88-24218
ISBN   0-89006-330-3

Copyright ©   1988
ARTECH HOUSE, INC.
685 Canton Street
Norwood, MA 02062

International Standard Book Number: 0-89006-330-3
Library of Congress Catalog Card Number: 88-24218

10   9   8   7   6   5   4   3   2   1

Contents

Preface

This book addresses the advantages and disadvantages of using CENTREX (central exchange) services to provide telephone and data communication facilities. Information about CENTREX and the experiences of its users are of vital concern to telecommunication and information systems managers in most businesses, institutions, and governmental organizations in the US and Canada.

Basic CENTREX was first introduced in North America more than 20 years ago, in large metropolitan areas, but became somewhat less popular in the late 1970s. Now that a large number of digital central office systems have been installed, CENTREX is experiencing a significant resurgence. Some experts believe that 20% of all business telephone lines will be provided through CENTREX within five years; CENTREX thus represents a serious competitive threat to private branch exchange (PBX) and key systems suppliers.

We have included case studies to illustrate the use of CENTREX in financial, retail, and government services. These case studies are favorable to CENTREX in that the organizations have decided to use it in preference to owning PBX systems, at least for some of their locations. We must emphasize, however, that we can easily find two organizations that decided against CENTREX for each one that opted for it.

To avoid repetition in this book, much generally useful information about CENTREX inherent in the six case studies (Chapters 7, 8, and 9) is not repeated elsewhere. This means that each of the case studies may be read by anyone interested in the advantages and disadvantages of CENTREX, even if the reader is not personally concerned with the specific applications described. The costs quoted in these case studies are in either Canadian or US dollars, as appropriate to the location of the system described.

ACKNOWLEDGEMENTS

We are grateful to the consultant liaison manager at each major telecommunication equipment manufacturer for supplying appropriate information, and to representatives of several telephone companies for their cooperation.

The following managers are thanked for contributing their time and interest through in-depth, face-to-face interviews that resulted in the case study reports: Adrian Berry, T. Eaton Company, Toronto; Evie Caldwell, Iredell County, Statesville, NC; Bill Dinwoody, Royal Bank of Canada, Toronto; Glen Houghton, Wells Fargo Bank, San Francisco, CA; Judy McMahon, Toronto Dominion Bank, Toronto; Wayne Salmon, Citibank Canada, Toronto; Michael Toner and Jody McCann, State of Wisconsin, Madison, WI.

Chapter 1
CENTREX in Review

1.1 INTRODUCTION TO CENTREX

CENTREX (central exchange) is the generic name for the provision of tele-communication services similar to a private branch exchange (PBX) from a central office, rather than from equipment on the customer's premises. In other words, CENTREX is a shared telecommunication service provided by the local telephone company that owns and operates the public telephone system. With CENTREX the facilities of a large central office system are logically partitioned by software and rented to a number of business customers as virtual PBXs. The central office is not dedicated to CENTREX service, but may be shared by a number of CEN-TREX subsystems, and is also used for regular local telephone service by businesses and residences. This situation is illustrated simply in Figure 1.1.

The numbering scheme used with CENTREX service is such that each local is given a conventional seven-digit number, but only the last four or five digits are used for internal dialing. Outside callers are encouraged to use the specific number of the desired extension, rather than dialing the switchboard number so that incoming calls bypass the attendants and go directly to the intended destination number. It is common practice, but not mandatory, for all of the lines on a CEN-TREX system to have the direct inward dialing (DID) feature.

1.2 DEVELOPMENT OF DIGITAL CENTREX

CENTREX service has been available in the major metropolitan areas of Canada and the US since the early 1960s. CENTREX was originally provided from step-by-step exchanges and was later available on crossbar exchanges. The main appeal of early CENTREX systems appeared to be for large, downtown businesses because its real advantage was that heavy, space-consuming switching hardware was in a nearby telephone company equipment building, rather than in the customer's office. Electromechanical CENTREX service needed bulky mul-tipair cables, similar to those used in the old electromechanical key telephone systems, in order to provide any features other than plain old telephone service (POTS).

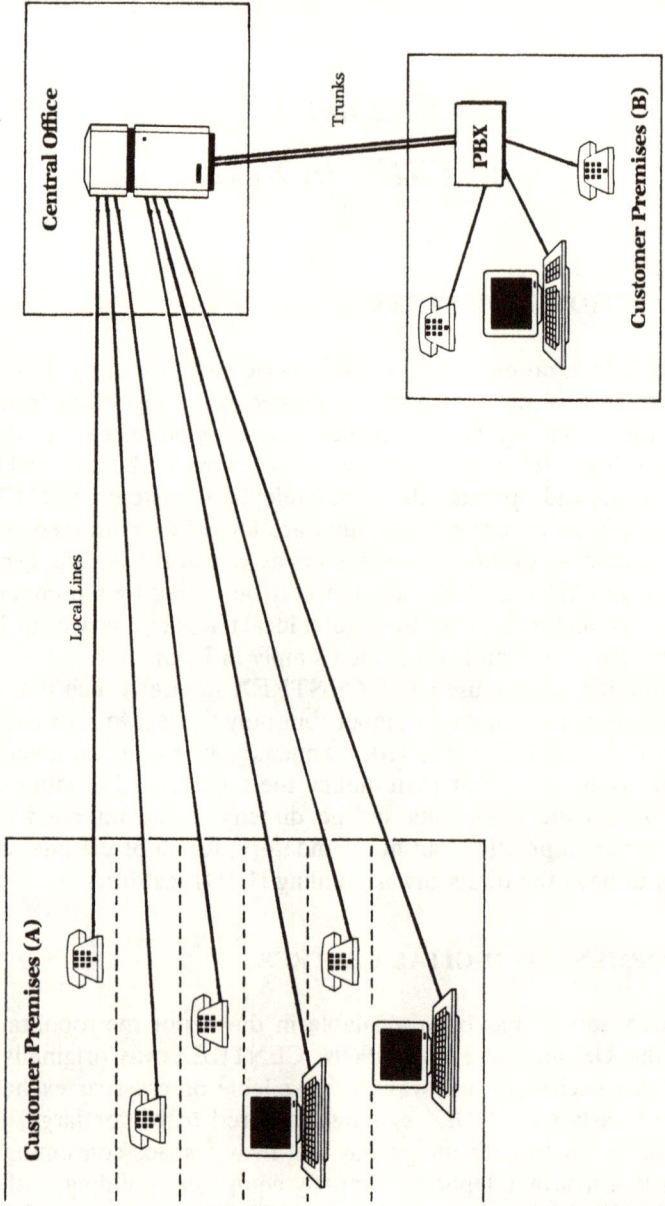

Figure 1.1 CENTREX or PBX.

Modern CENTREX services first became available in 1984, based on fully digital i.e., computerized, exchange systems, such as the AT&T No. 5 Electronic Switching System (5ESS), Northern Telecom Digital Multiplexed System (DMS-100), Siemens (EWSD), and GTE (5EAX). Following the divestiture of the Bell operating companies (BOCs) in 1984, the senior management of AT&T at first downplayed CENTREX and emphasized their new digital System 75 and System 85 PBXs. However, under the pressures of competition from Northern Telecom (which was developing CENTREX III in cooperation with its parent company, Bell Canada) and from the 22 former BOCs, which wanted to improve their marketing position, AT&T changed its emphasis fairly quickly. By early 1985 AT&T was offering its own version of digital CENTREX on 5ESS central offices to the BOCs. The enthusiasm for the marketing of CENTREX is still quite uneven in the US. The president of one BOC was quoted in 1986 as stating that "CENTREX is a dog that won't hunt." CENTREX III is now being sold aggressively by Bell Canada and its associate companies in eastern Canada.

Digital CENTREX is delivered to the desk over "skinny" wiring (one to three pairs) and offers nearly all of the many features that can be provided by a modern digital PBX. The service appeals especially to any organization that has a number of offices or other business locations in one urban area. Good examples of CENTREX customers are multibranch banks and retail chains with a number of stores and warehouses in a city.

The available estimates for the numbers of telephones on CENTREX in the US differ widely. Our best estimate is that there are about six million CENTREX-attached telephones in the US (i.e., about 10% of all business telephones). Some authorities quote a figure of nearly ten million CENTREX lines in the US, but this was probably the high point reached around 1978, before new digital PBXs began to erode the market for electromechanical CENTREX. Consultants to the telecommunication industry seem to agree that there is now a resurgence of business and the number of CENTREX telephones in the US is growing by at least 5% (i.e., about 300,000 lines) per year. This compares with shipments of about four million locals on PBX systems in 1986, where the five largest suppliers hold 80% of the market. By mid-1988, Northern Telecom had three million lines of Meridian Digital CENTREX installed or on order in North America. This means that Northern has supplied over 70% of all digital CENTREX lines up to that time.

New York Telephone claims to have installed 90,000 lines of digital CENTREX (known as Intellipath II) in 1987, of which 20% replaced PBX systems and the balance were conversions from analog CENTREX. During the same year a number of large customers in New York moved from analog CENTREX to their own in-house PBX systems.

Over 50% of all CENTREX lines in the US are still provided from analog central offices, primarily the 1A ESS. Since the telephone companies wish to obtain a reasonable economic lifetime for these systems, it seems likely that some analog CENTREX will remain in use until 1995.

In Canada there are about four million business telephones, of which two million are on PBX systems and more than one-half million lines are provided by CENTREX services. The other 1.5 million telephones are on single business lines or an electronic key telephone system. The Canadian federal government has the largest single CENTREX III system, with some 80,000 telephones in more than 100 office locations in the Ottawa-Hull national capital region, based on several DMS-100 central systems. Four of the five major Canadian banks use CENTREX III in the three biggest cities, probably totaling nearly 30,000 lines for these four customers. Some 13% of all business telephone lines in Canada are now provided through CENTREX.

1.3 ATTRACTIONS OF CENTREX

The advantages of using CENTREX service, as compared with an in-house PBX system, can be summarized as follows.

(a) There are significant savings on switchboard operator positions, with the accompanying cost savings, because most incoming calls go directly to the destination extension. In this sense, using CENTREX is similar to having a PBX where all the extensions are provided with direct inward dialing. For example, the head office of a major Canadian bank (Toronto-Dominion) requires only three attendants in the busy hours, compared with the need for at least six attendants on a comparable 3600-extension, modern PBX.

(b) All the maintenance of the telephone system, the additions, moves and changes, and the upgrades of software and hardware, can be performed by technicians from the telephone operating company (telco). This minimizes the need for customer-employed technical and management personnel. The telecommunication group can be kept very lean, compared with a similar department in a company that plans, manages, and maintains its own PBXs. This arrangement also provides the security blanket of 24-hour coverage for trouble calls and the expectation that the telephone company will keep the system's software and hardware completely up to date.

(c) Because the switching equipment for CENTREX is in the telco's premises, there are worthwhile savings on the costs of floor space, air conditioning, and electrical power. The elimination of the need for a large equipment room can be a major advantage for customers who are in expensive city center office buildings.

(d) One CENTREX system can now provide citywide service to offices in many locations, with a common numbering system, centralized attendant positions, and feature transparency. This gives real savings on private tie lines that would otherwise be needed and on operating personnel. A common system is also much easier to use, both for callers from the outside (e.g., customers) and for company personnel.

(e) There is practically no limit to a smooth expansion of the system size, up to well over the capacity of any single PBX in one urban area for one organization. The lower limit of size for the use of CENTREX is set by local regulations and tariffs, rather than by any technical consideration. In addition, we should note that mixed CENTREX-PBX private networks are possible using several of the common signaling systems.

1.4 MARKETING OF CENTREX

In many cases CENTREX services are sold by the same organization, and in some situations by the same salesperson, that is selling, leasing, or renting PBX systems. This may certainly be true in the US where the regional BOCs and the GTE companies provide CENTREX III (or its equivalent) and sell a variety of PBX products. Since November 1986 the BOCs have been allowed, by the FCC, to sell CENTREX services and customers' premises equipment (CPE) directly, rather than through separate subsidiaries. This was confirmed by an FCC order on December 30, 1987.

Digital CENTREX is sold under a confusing number of trade names. Southern Bell calls the service ESSX, while Southwestern Bell uses the name PLEXAR. Some Bell companies label this service as CENTREX III, while others refer to it as CENTREX IV. In western Canada the name Centron is used.

In many areas of the US the BOCs have also appointed dealers who sell CENTREX services, together with PBX and electronic key telephone systems, on their behalf. Because these are often nonexclusive arrangements, there may be several sources for CENTREX in some large urban areas.

In Canada, Bell Canada has folded up its PBX sales subsidiary (Bell Communications Systems Inc.) and now provides all of its business systems, including CENTREX, through one division — Bell Information Systems.

This situation means that the CENTREX and PBX sales teams may have a significant conflict of interest with providing the optimum systems solution for any given customer. If excess CENTREX capacity is available in an area then that service may well be emphasized and attractively priced to the detriment of the PBX solution. The telecommunication manager and other decision makers with potential customers need to be fully aware of this situation.

We can also see from the recent advertising from AT&T and Siemens that some "selling through fear" has developed. The suggestion is that CENTREX users will be able to take full advantage of developing integrated services digital network (ISDN) services much more easily and quickly compared to those organizations that have their own in-house PBX systems.

The pricing per line for CENTREX telephone service seems to be determined by the telephone operating company more to compete with the equivalent cost of a PBX extension rather than on the basis of the true cost of providing the CENTREX system. Senior executives from AT&T were quoted, in 1986, as stating that "CENTREX is a goose that lays golden eggs" and that "CENTREX, a long-time winner, looks even better now." It is very hard for an outsider to determine whether CENTREX tariffs are fully compensatory (or even over-compensatory) to the telephone company.

We do know that the per line capital cost of large, digital central office systems is lower than that of medium-sized PBX systems. In addition, the costs of selling one large central system to a telephone company may well be less for the manufacturer, such as AT&T, Northern Telecom, or Siemens, than the costs of selling one hundred PBXs to that number of different customers. As one example, the latest annual report of Northern Telecom indicates that the DMS central office product line brings in about 53% of its gross revenue compared with more than 60% of net profit. These numbers mean that CENTREX appears to be an excellent business proposition for both the telecommunication equipment manufacturer and for the telephone operating company. Whether this is an "all win, no-lose" situation that is also best for the third party, the customer, is not easy to decide.

With modern technology the minimum size of package that can be offered on CENTREX is falling. This entry level is now set largely on the basis of competitive regulations or by business considerations. The percentage share of business telephones that is gained by CENTREX seems to be inversely proportional to the minimum number of phones that may be connected to CENTREX from one location. In some areas, where no minimum is set, CENTREX has gained up to 20% of the market, and where a minimum of 100 or more telephones on CENTREX per location applies, then the service gains not much more than 10% of the market.

As the population of digital exchanges increases, then CENTREX could, in theory, be provided in any location in North America very soon. Most telcos in the US have emphasized small business CENTREX packages, at the level of two to six lines, over the last year. This service is replacing older (1A2) key telephone systems, which are still being used in large numbers. At the same time, area-wide CENTREX is also becoming more widely available. As the proportion of digital central office systems increases and CENTREX software becomes available from all of the telecommunication equipment suppliers, these developments should cause the growth rate of CENTREX to accelerate.

The electronic telephone sets that are supplied with CENTREX are identical in function and appearance to the sets sold with the same manufacturer's PBX. Telephones with multiple line appearances, programmable and fixed features keys, alphanumeric display, and hands-free, loud-speaker features are available. Some of the telcos will soon be offering fully digital telephones with the ISDN versions of CENTREX. These digital sets will make it less expensive and more convenient to attach data workstations through a CENTREX system. A number of electronic key telephone systems that provide enhanced capabilities are specifically suitable for installation with CENTREX systems. Two or three vendors have also announced specialized processors to provide a full range of office automation functions, such as voice and electronic mail, word processing, and spreadsheet packages, in conjunction with a CENTREX system.

1.5 DATA THROUGH CENTREX

Because modern CENTREX service is based on fully digital switching systems, providing a standard, two-way digital channel for voice communications, CENTREX systems are able to switch and transmit asynchronous and synchronous data streams, regardless of code or protocol, up to bit rates of 64 kb/s.

Most management information systems (MIS) managers and electronic data processing (EDP) managers seem to be more worried about the security of their data communication than telecom managers are about voice traffic. The concept of sending data streams across several city blocks and through a public switch, that may well be physically shared with a competitor, to achieve the equivalent of an in-building local area network (LAN) does not appeal to many who make decisions about data processing systems. At present it is difficult enough to convince an IBM-oriented MIS/EDP manager to rely on simple twisted-pair wiring as a replacement for expensive coaxial cables between terminals and the cluster controller. The idea of transmitting some data through a CENTREX network may well make economic and operational sense for an organization that is scattered across a city, but it is probably pointless for most potential users who are located in the same office building as the corporate mainframe computer.

This subject is considered in more detail in Chapter 4.

1.6 SOME CONCERNS WITH CENTREX

It is a common saying that "you cannot have the advantages of life without the disadvantages." That dictum is certainly true in the case of CENTREX, where we can identify one concern for each attractive feature.

We would summarize the major disadvantages of CENTREX as follows:

(a) The telecommunication department necessarily has less control over a CENTREX-based system than over a PBX (or a network of PBXs). Most telcos now offer a systems administration subsystem for customers to program their own moves and changes and to obtain call detail information. These are extra cost services, usually charged on a per line basis. The scope of these services is defined by the telco and will give less flexibility than a situation of complete system ownership.

(b) The rental costs of each voice line and data connection on CENTREX are subject to tariff regulations and will inevitably rise every year, at least by a few percentage points. Rate stability contracts that run for several years are common practice in North America and should be seriously considered for these services.

(c) There are sure to be speed, security, and protocol limitations in providing data communication through CENTREX. These must be determined and accepted by the MIS management of the organization.

(d) The customer may own the in-building telephone wiring but will not own the main distribution frame or concentrators used with CENTREX systems, as compared with complete system ownership in the case of a purchased, or leased, PBX.

Many offices are now using, or planning to use their existing twisted-pair wiring to replace coaxial cable, to provide RS232C or RS422 data links, and for local area networks, such as IBM's Token Ring Network (at speeds up to 4 Mb/s) and AT&T's Starlan or ISN. This multiple use of an integrated wiring system is not possible if the telco owns the cables for the sole use of CENTREX services.

(e) CENTREX service may not be available everywhere and this may inhibit the operation of private CENTREX-based networks already in use and the development of new networks.

1.7   EVALUATING CENTREX

The attractions and concerns of CENTREX service that we have listed, in Sections 1.3 and 1.6, are the main criteria for an intelligent evaluation of CENTREX for a specific organization, whether it be a business, institution, or government department, against the option of customer-owned PBX(s). A comparison of the advantages and disadvantages of CENTREX, and two sample financial evaluations are given in Chapter 6.

CENTREX can offer significant cost savings and productivity improvement under the following five areas: personnel (operating, technical, and management); space; private lines; management time; and ease-of-use. On the other hand the five main concerns that CENTREX customers face are: some loss of control; possible cost escalation; limited data communication feasibility; fewer opportunities for full system integration; and the potential lack of CENTREX service for some business locations. We estimate that for up to 20% of all business telephone users the five positives will outweigh the five negatives. These advantages and potential disadvantages are summarized in Table 1.1.

Table 1.1 CENTREX: Positives and Negatives

CENTREX offers	CENTREX may result in
Personnel savings	Loss of control
Network cost savings	Possible cost increases
Less management effort	Restricted data communications
Ease-of-use	Limited systems' integration
Space and power savings	Hybrid CENTREX-PBX network

The question of CENTREX *versus* the PBX is much more than a decision about possible dollar savings, but goes to the heart of corporate information systems policy regarding the degree of control that is desirable. The CENTREX user has handed back part of the organization's control over its future development to an outside vendor. In one sense, to choose CENTREX is a return to the comforting, but also restricting, embrace of the telephone company, from which legislation has released us over the last few years.

Chapter 2
CENTREX Systems and Features

2.1 VOICE SYSTEMS AND SUBSYSTEMS

This chapter is concerned with the voice (telephony) aspects of CENTREX systems. The central office systems that are available to support digital CENTREX services in North America are reviewed to help identify the similarities and differences between these systems. The reviews are in alphabetical order of manufacturer (i.e., AT&T, GTE, Northern Telecom, Siemens, and Stromberg-Carlson).

Three other manufacturers now also market central office switches into the North American marketplace. L.M. Ericsson has sold a few of its powerful AXE digital switches to Bell operating companies. Ericsson has been awarded a contract by British Telecom for AXE-10 switches to support 250,000 lines and it was stated that CENTREX software was included in that contract. However, Ericsson has not announced any plans for CENTREX in North America. NEC has sold several digital central office switches (the NEAX 61E) to at least one Bell company in the US. Adjunct switches that are based on the NEAX 61E have also been installed alongside existing analog central office systems to support ISDN services. No information has been given by NEC about the possibility of CENTREX service on the 61E system even though CENTREX and ISDN services will, obviously, converge within a few years. Mitel has announced the GX-5000 (derived from its SX-2000 PBX) as a central office system for suburban and rural locations. No plans for CENTREX have been announced for that system at this early stage.

We realize that a customer has no control over the specific central office switch that is used by the telephone operating company in any area. However, because there are differences in capability and in the level of development from system to system, it is worth knowing what is being offered by the local monopoly.

With older CENTREX systems it was fairly common for customers to install key telephone systems in conjunction with CENTREX to provide enhanced capabilities and, possibly, to save money by sharing CENTREX lines between a larger number of telephones. Some manufacturers now offer electronic key telephone systems (EKTS) that are designed especially to enhance CENTREX. This topic is addressed in Section 2.8.

There are at least 100 features that are available with CENTREX offerings. In Section 2.9 we classify the voice-related features into three categories and attempt to identify those features that are more common and most useful.

2.2 AT&T'S CENTREX-ISDN

AT&T started selling CENTREX on its fully digital No. 5 Electronic Switching System (5ESS) in 1986, under the name of Business and Residence Custom Services (BRCS). The 5ESS is being made by AT&T at the rate of about eight million lines per year and it is now, by a small margin, the most widely used digital central office system in the US. BRCS enables different packages of CENTREX features to be delivered to large and small businesses and to residences from the one model 5ESS.

2.2.1 Architecture

The 5ESS central office switch has a distributed processing architecture, so that the processing capacity is distributed across different 32-bit processors in the system. A simplified block diagram of the 5ESS is in Figure 2.1.

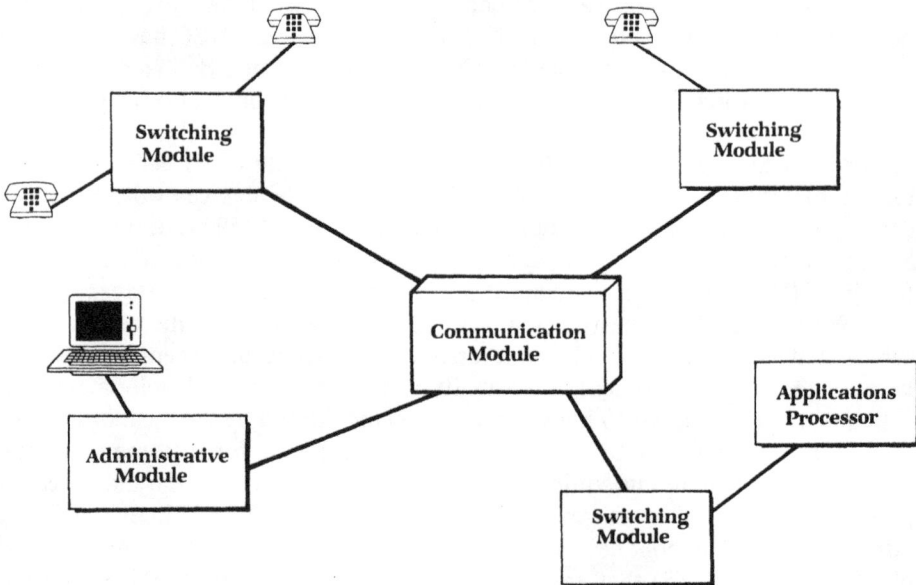

Figure 2.1 Basic Architecture of 5ESS.

2.2.2 Capacity

The communication module (CM) of the 5ESS can support up to 192 switching modules (SM), through fiber optic links. Multilevel space and time division switching in the 5ESS is done in a time multiplexed switch (TMS) within the CM. The switching network's capacity is determined mainly by the TMS, which can be expanded to accommodate the expected traffic volumes to provide an acceptable probability of blocking. The SM processor handles most of the call-processing work in the 5ESS and each SM has a rated call capacity of about 10,000 calls (of the simplest type) per hour. In practice the network is designed to limit the number of calls per SM to a much lower level. Up to 150,000 lines could be connected to one fully equipped 5ESS.

The administration module (AM) provides a sustained call capacity of over 300,000 *call completions* per hour, which AT&T claims is equivalent to around 600,000 busy hour call attempts (BHCA). This is the criterion more commonly used to compare switching systems.

A variety of peripheral interface units is available for the 5ESS, including support for groups of digital lines and digital trunks, as well as the full range of analog circuits. AT&T has an Integrated Services Line Unit (ISLU) that can have either of the U or T interfaces for ISDN applications and has a fully equipped capacity of up to 512 lines. The ISLU provides interfaces to both analog and digital line terminations and may be used in local or remote versions.

The applications processors that may be associated with CENTREX services running on the 5ESS are summarized in Chapter 3.

2.2.3 Remote Modules

Because a digital CENTREX service may need to serve a widespread suburban or semirural area it is important to have a number of remote module options. Five of the possibilities with the 5ESS are in Figure 2.2.

The remote switching module (RSM) can be equipped with software that gives it the capability for autonomous local switching if facilities at or between the host 5ESS and the RSM fail. This means that in a CENTREX application an RSM (in a single or multimodule configuration) could be installed on a customer's premises and be available as both a remote concentrator and a local switch. One timing-sensitive limitation is that the RSM must be within 150 miles of the host 5ESS, to which it is linked by T1/DS-1 digital carrier.

The Integrated Services Line Unit is a peripheral to the switching module, to support analog, digital, and special line terminations. A remote ISLU may be located on the customer's premises and can terminate 512 analog lines.

The optical remote module (ORM) is functionally the same as the RSM (or local SM in a 5ESS), except that it uses duplicated standard optical fiber facilities

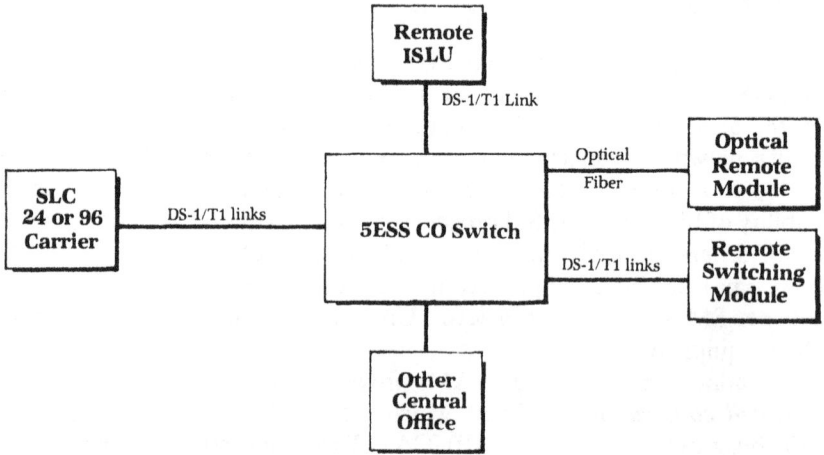

Figure 2.2 Remote module options with 5ESS.

to the host central office and it does not require a switching module at the central site, as the fiber optic link goes directly to the switching matrix.

AT&T has sold a large number of digital Subscriber Line Carrier (SLC) systems, which can be configured with 96 or 24 channels each. Integrated SLC-96 modules can be incorporated into a 5ESS host and these communicate over digital links with SLC-96 system remote terminals, which are located close to or within, the customer's premises.

2.2.4 Large Networks

The 5ESS, and hence the CENTREX systems that are resident in 5ESS offices, can be part of several different private switched networks for voice and data applications.

An electronic tandem network (ETN) is a telephone network in which a call can pass through intermediate switches before reaching its destination, as illustrated in Figure 2.3. The 5ESS can be either an end switch or a tandem switch in an ETN. The ETN software can also be based in the AT&T System 75 and System 85 PBX systems. In addition, ETN is compatible with the Meridian Switched Network (MSN, formerly known as ESN) package that is provided by Northern Telecom, so that mixed AT&T-Northern CENTREX-PBX networks are possible. ETN and MSN use channel-associated signaling and do not yet, therefore, take advantage of the sophisticated, high-speed signaling capabilities that will become available through common channel signaling and ISDN.

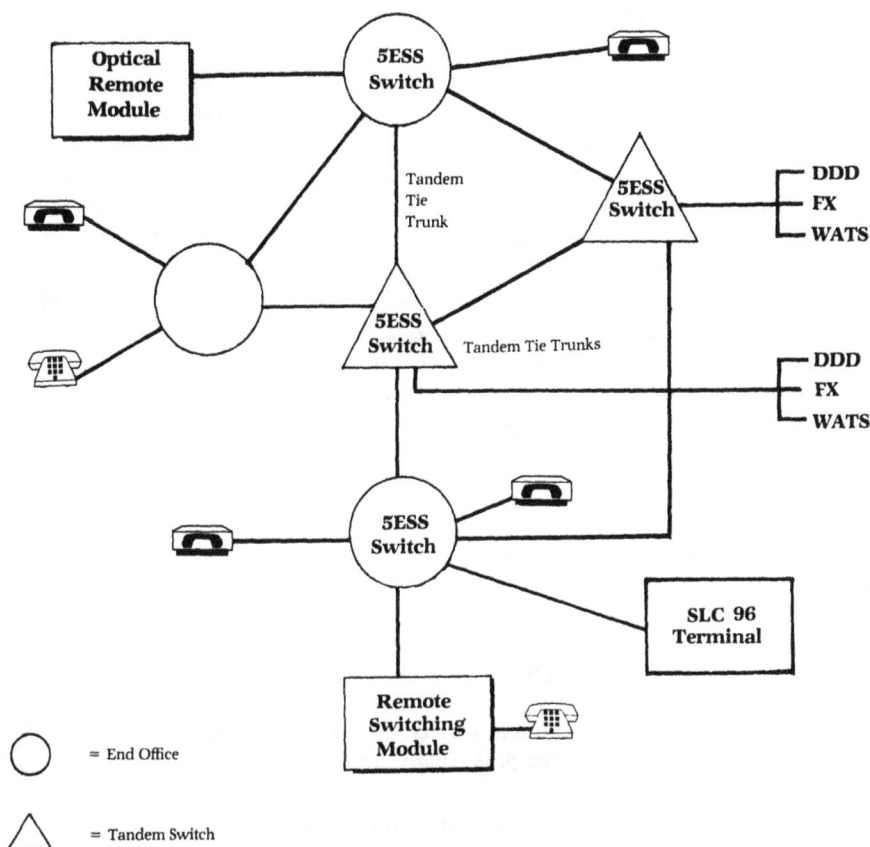

Figure 2.3 Possible electronic tandem network.

An ETN, linking CENTREX or PBX systems, can have a uniform numbering plan across a network. It provides most of the more significant features that are useful within a single CENTREX system, such as automatic route selection, message detail recording, and network class of service codes.

A 5ESS can also be part of a Common Control Switching Arrangement (CCSA), which is an older, private switched network offering from AT&T. CCSA provides a smaller subset of features within a network than does ETN.

A business customer on 5ESS-based CENTREX can subscribe to the software defined network (SDN) service. A user of SDN can build a specific and dynamic network within AT&T's switched network, including 800 and WATS services, to maximize efficiency and effectiveness. This management of SDN can be done through an on-line terminal in the customer's premises.

2.2.5 Citywide Networks

AT&T will soon be able to offer virtual citywide CENTREX service across multiple central office locations. This type of metropolitan area network (MAN) requires the availability of a common channel signaling (CCS) network and major software changes that will be part of the 5E5 generic (software release). For customers that are completely served by one host 5ESS switch, possibly with remote modules (such as the ORM, RSM, or SLC), all of their lines can be treated as a single CENTREX system now.

Citywide CENTREX has five major advantages for saving costs and improving service for multilocation organizations. These include: the elimination of local tie trunks; centralized attendant service; uniform dialing plan; identical feature sets in all locations; and pooled, easy-to-use access to private services such as WATS and foreign exchange (FX) lines.

Even at the level of residential telephone service there may well be applications for citywide CENTREX, with the restriction that any one line can be a member of only one network. Such networks could be a valuable service for members of a business or a professional partnership or, for example, of a municipal council.

2.2.6 Customer Premises Equipment

AT&T has two console models for its CENTREX system—the 50B attendant's console, which is part of the 50B Customer Premises System, and the ISDN attendant's console.

The .50B can be obtained as a very compact unit, with an eight-character display for incoming call identification and time of day. An add-on panel provides 50 direct station selection buttons and a busy lamp field to display the busy or idle status of any 100 stations at a time.

The ISDN attendant's console has a larger alphanumeric display, on a smaller footprint, than the 50B and can have an add-on direct selector unit with 100 keys and lamps.

AT&T provides a wide range of analog and hybrid telephone sets and has demonstrated its first ISDN-compatible digital telephone, with an optional RS-232 or V.35 data port.

2.3 ADVANCED BUSINESS SERVICES (GTE)

GTE sells its GTD-5 EAX digital central office to the GTE-associated telephone companies and other independent telcos in the US, as well as to its two partly owned Canadian subsidiaries, British Columbia Telephone and Quebec Tel.

GTD-5 EAX systems have been installed at the annual rate of about one million lines for several years.

GTE offers a package of features known as Advanced Business Services (ABS) that equates to digital CENTREX service. At the moment ABS only applies to telephone (voice) service, although data communication facilities are promised for the future. GTE operating companies are participating in several major ISDN trials in the US. GTE emphasizes, in particular, that the GTD-5 was designed with ISDN in mind and that ISDN basic rate interface cards are plug-compatible with analog line cards.

Because the GTE companies generally serve suburban areas and have a smaller total customer base than the smallest of the regional BOCS, there is far less CENTREX experience in this company than with AT&T or Northern Telecom.

2.3.1 Architecture

The GTD-5 EAX is based on modular, distributed architecture, with full redundancy throughout the switching network, the central control, and all common peripheral equipment. This is the only system with fully duplicated digital terminations.

Packet-switching capabilities are integrated within the GTD-5, rather than being provided by another attached processor. This part of the GTD-5 architecture corresponds to that of the OMNI PBX family, that was formerly made and sold by GTE. This feature should mean that the provision of X.25 network access for data communication should be only a marginal cost.

2.3.2 Capacity

The theoretical capacity of the base unit in the GTD-5 system is 150,000 lines or almost 25,000 trunks, although it is doubtful whether any system would be taken beyond 100,000 lines. In CENTREX (ABS) service a single system can support up to 30,000 lines. The main memory can be expanded to 96 million bytes.

No specific figures on the BHCA power of the GTD-5 have been published, but we can assume that this switch is roughly comparable in processing power to the AT&T 5ESS.

2.3.3 Remote Modules

GTE offers four remote units with the GTD-5. These are illustrated simply in Figure 2.4.

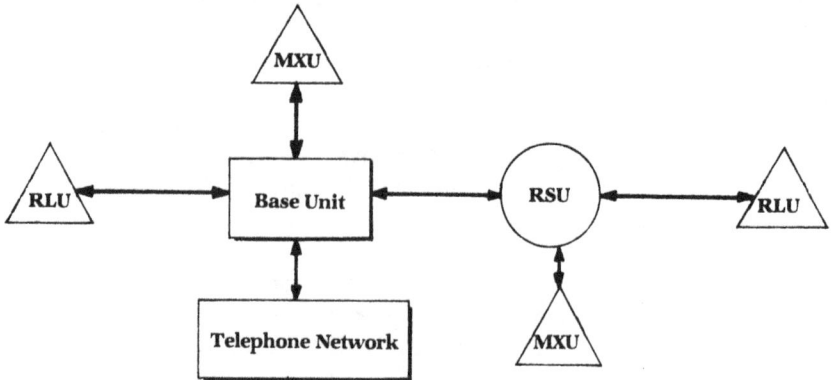

Figure 2.4 GTD-5 EAX with remote units.

The remote switching unit (RSU) can handle up to 6,000 telephone lines, which are linked and controlled from a base unit (BU). The RSU can maintain local service if the base unit link is broken, but it is probably too large to be installed in a typical customer's premises.

There are two line-concentrating modules. The remote line unit (RLU) can connect, with T1/DS-1 links, to a BU or an RSU and can support up to 1536 telephones. The RLU can be converted to an RSU (i.e., be given switching capabilities) without being taken out of service. The MXU is a multiplexer unit that provides up to 192 lines, preferably over a fiber link.

GTE has been active in optical fiber transmission and switching developments and can provide both fiber and copper-based digital transmission systems for the various GTD-5 units.

2.3.4 Networks

The GTD-5 is compatible with the ETN and CCSA that were originated by AT&T and can thus be part of a nationwide telephone network. Older designs of tandem tie trunk networks can also be linked into the GTD-5.

Citywide CENTREX networks can be built, for up to 30,000 phones, from one BU with a number of RSUs and have complete feature transparency. In this type of configuration, with an RSU some miles away from the host GTD-5, the business survivability of the RSU becomes important. That is, can the RSU provide basic business service on its own?

2.4 MERIDIAN DIGITAL CENTREX (NORTHERN TELECOM)

Northern Telecom was the first major telecommunication manufacturer to emphasize fully digital switching systems. It sells a family of digital multiplexed systems (DMS). The most important switch in this range is the DMS-100, which is a close second to AT&T's 5ESS in the US market, dominates the Canadian central office scene, and is used in several western European telecommunication networks. The DMS-10 is intended for smaller installations and the DMS-250 is designed specifically as a tandem switch, for which purpose it is used by a number of independent long-distance carriers, such as MCI and CN/CP Telecommunications.

In recent years the DMS products have been manufactured at an annual rate of about five million lines. Northern Telecom began shipping its digital CENTREX package to telephone operating companies in 1984 and installed its one-millionth line in mid-1987, with more than another one million lines of Meridian Digital CENTREX on order at that time. The company has an ambitious program to maintain its leadership in digital CENTREX, which it sometimes describes as Meridian Business Services. Northern delivers two or three major software upgrades, known as batch change supplements (BCS), each year, with about ten added features in each new BCS.

2.4.1 Architecture

Northern Telecom announced a major improvement to the DMS architecture in late 1987, in the form of DMS SuperNode. The improved call- and signal-processing capabilities of this multiprocessor configuration make the DMS competitive with the most powerful central office systems on the market. The main components of the DMS SuperNode are illustrated in Figure 2.5, which shows that Northern Telecom has effectively converted from a centralized, to a distributed architecture.

The DMS Core is a fault-tolerant computer that communicates with other subsystems through the DMS bus. The high throughput of the core-bus combination will become essential as ISDN and common channel signaling are widely used. The link peripheral processors (LPP) provide the interfaces to various analog and digital networks. Signaling System No. 7, both basic and primary rate accesses for ISDN, and X.25 packet-switching accesses are supported through multiple LPPs on the DMS Bus.

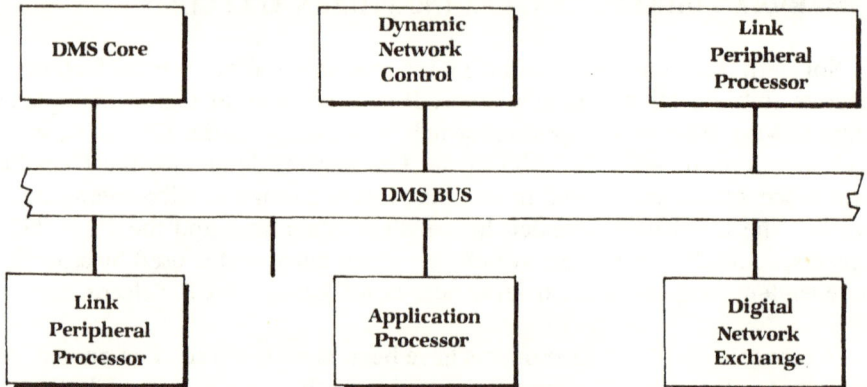

Figure 2.5 DMS SuperNode system architecture.

High-volume database services can be implemented on the applications processors. The Digital Network Exchange (DNX) facilitates the control of multichannel digital links, such as T1 and T3. The Dynamic Network Controller (DNC) subsystem can include custom-tailored application programs that may be written by telephone companies or large customers for specific market services.

These developments mean that Meridian Digital CENTREX can be provided on a networkwide basis and also can be configured to match the needs of a given customer.

2.4.2 Capacity

Because the DMS-100 was one of the first fully digital central office systems to come on the market, it was almost inevitable that later competing arrivals would offer improved performance. The announcement of DMS SuperNode should end any concerns regarding the throughput capabilities of this system. A planned series of upgrades to the DMS core will take its processing power up to 1.5 million BHCAs. The DMS bus has a message handling capacity of about 100,000 transactions per second. The applications processors on the DMS bus will each be able to approach 1,000 transactions per second. Taken together, these subsystems enable the DMS-100 to be used as a signal transfer point in a CCS7 network and also as a service control point for network services that need access to large data bases, such as advanced 800 service and on-line electronic directory services.

2.4.3 Remote Modules

Northern Telecom has several types of remote modules that are suitable for installation within a building or as an outside plant. Essentially, there are three different sizes of remote modules, as follows:

Remote subscriber line equipment = 1024 lines over 8 DS-1 links

Remote line-concentrating module =   640 lines over 4 DS-1 links

Remote subscriber line module   =   256 lines over 2 DS-1 links

These remote modules may be equipped with the emergency stand-alone option, which means that intrasite calls, where both ends are attached to the one remote module, remain connected even if communication is lost with the host DMS. The multichannel digital links (T1/DS-1) between the remote modules and the DMS-100 may be carried on four-wire copper circuits or on optical fiber links.

A typical CENTREX configuration with remote modules is illustrated in Figure 2.6.

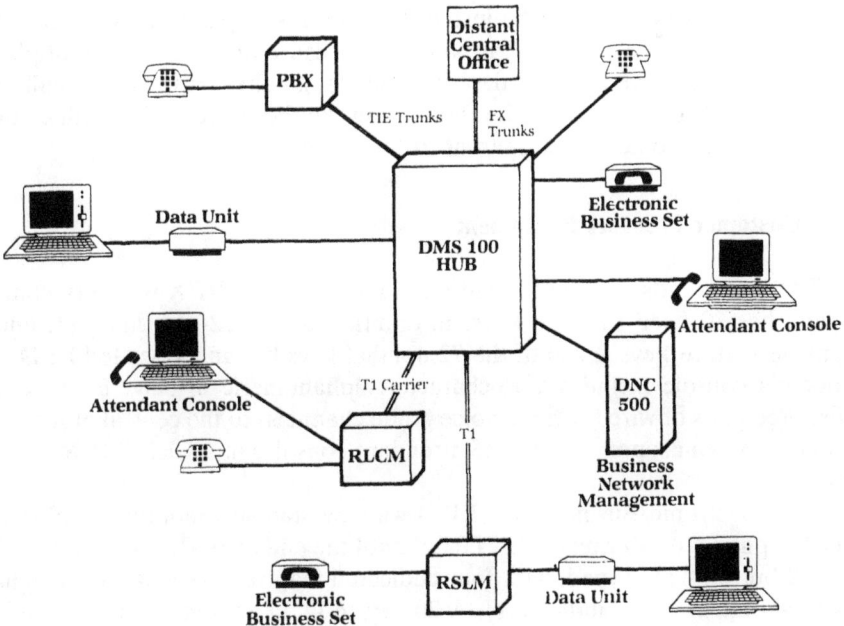

Figure 2.6 DMS-100 hub for Meridian Digital CENTREX.

2.4.4 Large Networks

Meridian Switched Network (formerly ESN) can be used in a CENTREX-only, a mixed CENTREX-PBX, or a PBX-only environment, with the MSN software running on DMS-100 and SL-1 switches. Most MSNs have so far been implemented by companies that rely solely on SL-1 PBX systems, but we expect that an increasing number of such networks will include CENTREX systems in the larger cities. MSN may also be used to effectively provide a single digital CENTREX network across a large urban area, so that parts of that network may be based on two or more separate DMS-100 offices. Northern Telecom has also announced the availability of automatic call distribution (ACD) on CENTREX, which thus competes with a major and fast-growing application for PBX systems.

2.4.5 Citywide Networks

Two approaches to a citywide Meridian Digital CENTREX network are available. The more common solution has been to install remote modules (RSLE, RLCM, or RSLM) close to the various locations that are on the network and to link these remote modules to one DMS-100 host. The remote modules may be in telco or customer premises, or be installed as an outside plant. This solution may not be economical in a very large urban area, because of the cost of multiple T1/DS-1 links to the single host. As described above, an alternative is to install MSN in multiple DMS-100 locations. The choice between these alternatives will be made by the telco and should be transparent to the users.

2.4.6 Customer Premises Equipment

The attendant's console for Meridian Digital CENTREX is a very compact unit and has 30 fixed function keys, in addition to the 12-key dial pad, and 42 assignable feature keys. Each of the 72 nondial keys has an associated LED. The attendant's console includes a 16-character alphanumeric display. Each console needs three pairs of wire (or three voice grade channels) to the central office switch and one DMS can support up to 255 attendant consoles in a CENTREX environment.

The Electronic Business Set (EBS) was the standard multiline analog electronic telephone for Meridian CENTREX until the end of 1987. The standard EBS had a 12-button dial keypad, with four dedicated feature keys and nine assignable keys (eight having LCD indicators). A 20 key/16 LCD add-on module is available and up to three of these may be attached to one EBS. A two-line, 16-characters-per-line, alphanumeric display is another option on the EBS.

The EBS requires only a single wire pair to the host, or remote module, because the signaling information is carried on a data channel centering on 8,000 Hz (above the 4 kHz voiceband). The EBS has been extensively criticized as looking too bulky and being both space-consuming and expensive.

Northern Telecom has now released a new family of low-profile telephone sets that require less desk space than the EBS. The simplest of these is the M5009, which has nine feature keys. The M5112 has a built-in speakerphone and can have a 36-button add-on module. Other versions of these multiline sets with alphanumeric displays are becoming available during late 1988. Northern Telecom's first ISDN-compatible phone, known as the M2317 (or the T2317) has been used at a number of CENTREX-ISDN demonstrations and will be shipped on general release in early 1989.

The most common single-line telephone on the Meridian Digital CENTREX system is the Unity set, which is compact and inexpensive. Other analog DTMF telephones can, of course, be used on CENTREX lines.

2.5 EWSD ISDN CENTREX (SIEMENS)

Siemens AG claims to be the third largest manufacturer in the world of telecommunication systems (after AT&T and Alcatel). Siemens' strength is in the broad range of its activities across the complete spectrum of electrical and electronic products and systems. In 1986 Siemens took over most of GTE's telecommunication business outside North America.

Siemens participated in several early ISDN demonstrations and field trials with Bell operating companies, using its fully digital central office system, which is known as EWSD. Siemens, Ericsson, and NEC are competing fiercely to be accepted as the third supplier of large public switching products to the Bell systems. At this stage it would appear that Siemens is slightly ahead and it is the only one of these three major companies to have announced a CENTREX package. Although information has been published on EWSD ISDN CENTREX there is no indication that this software has yet been cut over into live commercial service for any customer.

2.5.1 Architecture

Siemens took the same approach to the EWSD central office as it used with the HICOM PBX and designed the system to support ISDN applications without any architectural changes. The system supports CCS7 and will thus be able to provide subsecond call setup times. The system has hardware redundancy to the level of any competitive public switching system. The architecture is somewhat similar to the 5ESS, in that most of the call processing is handled in a number of switching modules.

2.5.2 Capacity

The EWSD has been designed for growth up to 100,000 lines on one system. Firm information on the BHCA capacity of the EWSD central office is not available.

2.5.3 Remote Modules

Siemens has a remote switching unit that can support both basic and primary rate ISDN accesses on the terminal side and is linked to the EWSD central office by T1/DS-1 or primary rate links.

2.5.4 Networks

EWSD CENTREX will support virtual networking capabilities for voice and data calls. By using CCS7, multiple EWSD systems can be linked through inter-office trunks to form a citywide CENTREX network. Since Siemens has not announced the availability of an equivalent to ESN or CCSA, it would appear that intercity networks based on the EWSD will require the availability of wide-area ISDN.

2.5.5 Customer Premises Equipment

The EWSD CENTREX attendant's console will be an IBM-compatible personal computer, linked to the host by a basic rate ISDN access. This console will provide menu-driven software packages for call control, traffic data collection, electronic directory, and station message detail recording.

Siemens is a large-scale supplier of telephone sets on a worldwide basis, to the extent that its telephones are used extensively on switching systems that originate from other manufacturers. Fully digital telephone sets, with the option of a data port, will be available on EWSD CENTREX from the start because of the system's compatibility with ISDN standards.

2.6 EXTENDED SERVICES PACKAGE (STROMBERG-CARLSON)

Stromberg-Carlson is a subsidiary of General Electric-Plessey Telecommunications (GPT), of Britain, and has been supplying its Digital Central Office (DCO) to telephone operating companies in the US and Canada for over six years. The DCO is most suitable as a small central office for suburban and rural areas.

It can be used as a large PBX on a campus and is successful as a switching center for cellular radio systems.

Over 850 DCO systems and 800 remote units are in use, for a total of 1.8 million lines. The DCO is now being made at a rate approaching 400,000 lines per year.

Stromberg-Carlson offers five levels of its Extended Service Package (ESP) for residential and business services. The fully featured versions of ESP are equivalent to CENTREX service on the larger switches.

2.6.1 Architecture

The DCO incorporates general purpose super minicomputers as the maintenance and administration processor and the call processors. Figure 2.7 illustrates the two identical switching systems, with one system actually on-line and the other in a hot standby condition.

If a failure should occur in either system, no calls would be lost, even those in the process of being switched. The call processors can also continue to operate independently if the maintenance-administration processor fails.

No announcement has yet been made regarding features or interfaces on the DCO specifically for data communication.

2.6.2 Capacity

The utilization of a telephone line (or switched link) is generally measured in hundred call second (ccs) units per hour. At a traffic load of 6 ccs per line and 28 ccs per trunk, the DCO can serve 16,000 telephone lines and over 2000 trunks. The processors provide a capacity of just over 100,000 BHCAs.

Theoretically, one Stromberg-Carlson DCO could have 100,000 directory numbers and 7500 trunks, but it is not realistic to plan on the basis of such a large single system.

The DCO is cost-effective as a central office in the range from 1000 to 32,000 telephone lines per system. It is also designed to be used as a tandem-only switch, in both common carrier and private networks, with up to 4000 attached trunks.

2.6.3 Remote Modules

A remote line switch is available for linking distant clusters of telephones to a DCO host. In addition, Stromberg-Carlson has introduced a compact DCO-SE package with a capacity of 1000 lines and 240 trunks. This cost-efficient configu-

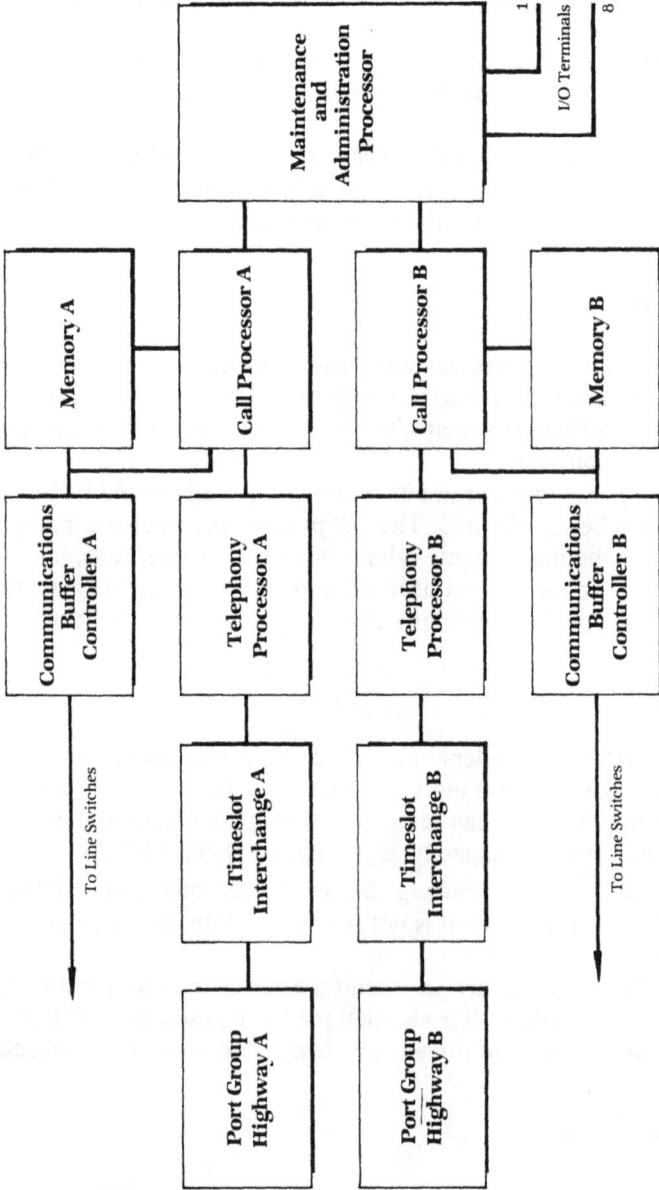

Figure 2.7 Stromberg-Carlson DCO.

ration can be used as a small host switch or as a remote module to a large central system. The DCO-SE could bring CENTREX-like telephone service to thousands of small communities that still have hard-to-maintain electromechanical dial offices.

2.7 VOICE MAIL

Voice mail is a user-controlled, centralized, digital answering and message-forwarding system and can be the most valuable and productive server attached to a CENTREX system. Voice mail is alternatively known as voice messaging, voice processing, or voice store-and-forward and, after a slow start in the market-place, it is now beginning to deliver the promised benefits. Voice mail systems not only store simple messages but can support call processing, work as automated switchboard attendants, and be used in a network with remote CENTREX or PBX systems.

A voice mail system allows the recording of messages by both the inside users of CENTREX and outside callers. The voice messages are stored in digital format on high capacity, fast-access disc drives. The use of voice mail can greatly reduce the frustration of "telephone tag" and cut down repetitive call attempts. Organizations that have a good proportion of their people away from the office, probably on sales or service calls, can especially improve communication, saving both time and money, with a voice mail system.

One serious consideration with a voice mail application is the relatively high cost of the subsystem, which is most likely due to the heavy storage requirements for the digitized voice signals. One minute of a voice conversation encoded at 64 kb/s would occupy nearly one-half megabyte. All digital voice messaging systems use compression techniques, typically reducing the storage requirements by up to a factor of four (i.e., to 16 kb/s). For example, one entry-level voice mail system is priced at about $50,000 for an eight-port system with 20 hours of storage. A relatively small system can support a large number of users, because the average recorded message is less than 30 seconds long and many users will call into the voice mail application outside regular working hours. An average purchase cost of $400 per user is probably a good figure to use for estimating purposes.

Some voice mail systems provide optional audible prompting messages to the occasional user and most give the user control over access to messages and to routing voice messages to multiple "mailboxes."

Voice mail is not yet being offered by all the telcos that provide CENTREX service. AT&T makes its own Audix system that can be closely integrated with the 5ESS. A system manufactured by VMX (which has supplied 25% of all voice mail subsystems on PBXs) is generally used in conjunction with the DMS-100. In these cases the voice mail system would normally be located on the customer's premises. At least one digital voice mail service bureau is now operating in most major cities in North America. Such a service could be used with CENTREX lines as easily as with other business telephone lines.

2.8 KEY TELEPHONE SYSTEMS

A significant proportion of customers on the earlier (analog) CENTREX services employed key telephone systems in their premises "behind" the CEN-TREX system. The purpose was to provide to selected users a range of features that were not available on the basic CENTREX service. Because no more than 30% of all CENTREX users are on digital CENTREX, there is still a worthwhile demand for electronic key telephone systems in this supplementary role. Clearly, this market for EKTS will decrease quickly over the next five years as most CEN-TREX users are moved to digital systems with a full range of features.

In the exceptional circumstance it may be financially worthwhile to install EKTS within a digital CENTREX system. If an office has a number of occasional telephone users, with a telephone utilization below 25% each, then the cost of a line on a purchased EKTS may prove to be less expensive than having a dedicated CENTREX line for each telephone. This could be the case in a small-to-medium sized installation, perhaps below one hundred lines, where the customer does not have the buying power to get the lowest possible price per CENTREX line, as compared with a large customer, with over a thousand lines. One good reason to consider EKTS in this cost-cutting role is that key systems are very competitively priced in a cutthroat market, whereas CENTREX is a regulated service that is provided by a local monopoly.

Small EKTS may be purchased from $150 to $400 per line and larger systems, up to as many as 100 lines, are priced in the range of $300 to $600 per line. In the CENTREX environment it is much more likely that a number of small EKTS would be purchased to be used on a departmental basis. If a 4-trunk, 16-line key system can be bought and installed for $4,800 it may be appropriate to spread the capital cost over 30 months and impute a monthly cost of $10 per telephone. If in the same installation, the cost of each CENTREX line is $20 per month, then a quarter of that cost (i.e., $5 per month) should be allocated to each telephone. In that example, the effective monthly cost per line on an EKTS behind CENTREX is $15, compared to $20 per line on the dedicated CENTREX system.

The use of EKTS behind digital CENTREX might also prove to be worthwhile if the tariff for certain desirable features is established at a somewhat high level. This second factor is becoming less likely as the software packages for CENTREX mature and as the telcos can offer an ever-widening range of features, particularly on the AT&T and Northern Telecom-based services. Because the S-bus with ISDN will be able to support up to eight stations on one line and CENTREX-ISDN will deliver a rich menu of features, we believe that the arrival of ISDN, whether as part of a CENTREX service or on a PBX, will be the death knell for supplementary EKTS equipment. It may well be the case that small CENTREX will compete very successfully over the next five years (in the range from $10 to $15 per line) against key systems in a wide variety of small businesses in urban and suburban locations.

Some suppliers have designed EKTS packages specifically for use in the CENTREX environment. A sample list of these EKTS vendors is given in Appendix A. This is certainly not intended as a complete list of all sources for EKTS in North America, because such a list would amount to at least forty original manufacturers, along with hundreds of dealers and distributors. It should be possible to use any EKTS that is designed to North American standards as a supplement to a CENTREX service.

2.9 FEATURES WITH CENTREX

Digital CENTREX services can now provide almost as many features as modern PBX systems, with one important consideration being that some of these feature packages are frequently extra-cost items with CENTREX and *may* be supplied as standard features on a digital PBX. For the time being the services based on Northern Telecom's switches provide a greater variety of features than do those based on the AT&T 5ESS, largely because DMS-based CENTREX was the first digital service on the market. With a major software release due before the end of 1988, it seems likely that the 5ESS will then offer users as many features as were available on the older 1A ESS and will significantly close the gap with the DMS-100.

The features provided through CENTREX, or on a PBX or EKTS, are generally classified into three groups. *System features* can be made available to all of the users on a system. *Station features* are provided on a station-by-station basis and may be dependent on the availability of a specific type of telephone set (e.g., one with hands-free operation or a digit display). *Attendant features* are those that are available to facilitate operation of the attendant's console.

The most frequently used and valuable features for the users of CENTREX are described briefly in the following paragraphs. A longer (but not complete) list of feature definitions is given in Appendix E. We should note that the terminology for features differs somewhat from one system manufacturer to another and that we have used the most commonly accepted wording here.

For detailed lists and descriptions of all the features that may be available on a specific central office system you should request the features description book for that system from the manufacturer.

2.9.1 Automatic Route Selection (ARS)

ARS makes it easy to ensure that all billable calls are connected through the least costly channels that are available at a given time. This system feature is usually associated with call back queuing, so that a user can be notified when a less expensive trunk becomes available.

2.9.2 Call Detail Recording (CDR)

CDR, frequently known as station message detail recording (SMDR), provides a centralized record of all calls made from every line on a CENTREX system. The combination of ARS and CDR may produce long distance savings of 20%, compared with an uncontrolled network.

2.9.3 Call Forwarding (CF)

CF allows a user to have calls automatically forwarded to any telephone on the CENTREX system, under a variety of conditions. Call forwarding–busy routes calls to another line when the user is on the telephone and call forwarding – no answer routes calls to an answering position after a prespecified number of rings. In some systems it is possible to forward internal and external calls to different numbers for answering purposes.

2.9.4 Conferencing

Conferencing usually allows a user to simultaneously connect up to 30 people in a group conversation. Conferences of up to six parties can be set up directly from a user's standard telephone and larger conferences can be established with the attendant's assistance.

2.9.5 Direct Inward Dialing

DID is a standard feature of CENTREX systems that allows incoming calls to be routed directly to the user's telephone, without having to be handled by a console attendant. This feature often reduces the attendant's workload by over 50% when compared with a switchboard not having any DID.

2.9.6 Hunting

Hunting can exist in two directions. Station hunting means that a system is configured so that incoming calls to a busy number are redirected to alternate stations. With trunk hunting, outgoing calls will search for an available trunk in the appropriate line group. A number of hunting procedures, such as circular hunting, uniform call distribution, preferential hunting, and queuing, are available within Station Hunting.

2.9.7 Speed Calling

Speed calling permits a user to dial from a list of frequently called numbers by using an abbreviated code. It is common to have one-digit speed calling for up to eight codes and two-digit speed calling for up to 30 codes. Speed calling lists may be shared by a number of users, but tend to be an expensive option on CENTREX, at several dollars per month for each 30-number list. It is worth noting that the name *speed* calling is incorrect, in our pre-ISDN systems, because the use of abbreviated dialing does not actually shorten the call setup time. The term *abbreviated dialing* is a more accurate description of this feature.

Chapter 3
Management of CENTREX Systems

3.1 MANAGEMENT PACKAGES AND SYSTEMS

There is clearly a trend with CENTREX services to provide a growing array of administrative and management tools to the customer. This trend is part of an attempt by the telcos to match, with their CENTREX service offerings, the features and facilities that come with digital PBX systems.

The management tools are presently available from two sources: the telco rents or sells services and systems to its CENTREX customers; and third parties sell systems that can be added to a CENTREX installation.

We can identify three major management applications needed with CENTREX: *system administration,* which looks after additions, moves, and changes (and may also include the attendants' consoles); *system management,* which is primarily concerned with call detail recording, cost allocation, and optimization; and *network management,* which is associated with problem determination and network configuration.

These three significant CENTREX management requirements are addressed in more detail, with some examples, in the following sections.

3.2 SYSTEM ADMINISTRATION

A prerequisite to a satisfactory arrangement for customer-controlled administration of a CENTREX system is to have a structured, customer-owned, telecommunication cabling system. This cabling system should provide jacks with a sufficient number of twisted wire pairs wherever a voice or data terminal is ever likely to be placed. The objective of such an installation is to plan for a system lifetime of ten years and to be able to handle all of the physical additions, moves, and changes without extensive recabling. This type of approach requires a fairly high investment in a well-planned cabling scheme, but can save several times its initial cost by minimizing new cable installation over the ensuing years. Another part of a well-planned cabling installation is the use of lockable wiring closets, provided on the basis of at least one per floor, to house all cable terminations and cross-connect arrangements together with any electronic hardware, such as controllers, that needs to be provided on a floor.

Several major suppliers now offer universal cabling systems that are based primarily on the use of twisted-pair copper wiring, avoiding the use of any coaxial or twin-axial cable, and that employ optical fiber wherever that may be justified. One of the best in-building cabling systems is the Premises Distribution Systems (PDS) from AT&T. The PDS includes patch panels and plug-ended cords for cross-connecting twisted-pair wiring and optical fiber cables. The AT&T 110 Patch Panel system provides for cross-connections of multiple circuits, in 300 and 900 pair cable sizes.

Northern Telecom offers its Integrated Building Distribution Network, which is a well-planned and fully documented system, also based on copper wire pairs and optical fiber.

A third alternative is the structured cabling system that was designed, and is still recommended, by IBM, now available from independent suppliers. The IBM cabling system is unique in its emphasis on shielded twisted-pair wiring, which can be valuable to ensure higher bit rates or lower error rates. Unfortunately the IBM-designed cabling system employs bulky plugs and sockets for its cross-connect panels and is usually three times as expensive to install as most competing systems.

A variety of other, similar, universal voice-data cabling systems is available from the telephone operating companies and from independent contractors. Regardless of the source of the wiring, this matter must be emphasized much more than it was in the days of simple analog CENTREX systems. It deserves careful design, planning, and installation, together with complete electrical testing of each of the circuits after installation and before being cut over for regular use.

The AT&T 5ESS offers on-site operations, administration, and maintenance (OAM) capabilities through an on-line terminal. Change and verification of the data base in the 5ESS may be done from the customer's offices, at the central office, or from a remote site. When this responsibility is shared by a system administrator employed by the customer and by telco personnel, there must be very close cooperation between the two groups. If the two OAM groups are not working in a coordinated way then it is better for the customer not to be involved in system administration at all.

Northern Telecom has announced its Customer Site Administration (CSA) package, although this has not been implemented as yet with many digital CENTREX customers. CSA is a computer-based subsystem that provides the customer with the ability to rearrange telephone sets and features. It requires a personal computer in the system administrator's office to interact with three modules in the DMS-100 at the central office, namely the Master Control Unit, Gateway, and Customer Assistance Center. The CSA package includes a system audit capability to keep a record of all adds, moves, and changes that are made in the data base, as well as a set of self-diagnostic reports.

The EWSD ISDN CENTREX system from Siemens will include a customer station rearrangement (CSR) capability that will be run from an administrative terminal on the customer's premises. CSR will be used to control the move of stations to different locations and to change the allocation of CENTREX features as needed. Various management reports will be available from CSR to provide the administrator with the necessary information to control the CENTREX ISDN environment.

Several independent companies now offer CENTREX management systems for administrative, as well as call-accounting, applications. One of the better known of these systems is Cenpac from American Telecorp, which is in use on systems with a total of over one-half million CENTREX lines. Cenpac is compatible with 5ESS and DMS-100 central offices (as well as with the older 1ESS and 1A ESS). This package automates line and feature changes, gives a user control of the CENTREX data base and provides a number of user-definable reports.

Other aspects of system administration that have been addressed by a few suppliers are the attendant's console and message center. For example, Conveyant Systems has announced the Teledesk CENTREX Workstation, based on a PC-AT. This unit provides enhanced call processing by substantially reducing the number of required key strokes; an electronic directory; and a message center.

Most telcos make an extra charge to users who employ a system administration terminal on their own premises. New England Telephone (part of NYNEX) has announced that with its Intellipath II (digital CENTREX) service in Massachusetts there will be no cost to use its Customer Line Administration feature.

3.3 SYSTEM MANAGEMENT

System management is usually associated with the major function of call accounting, which involves the collection of station message detail records from the CENTREX system and the processing of SMDR to produce call detail records. The CDR output can be distributed to departments or to individuals, to highlight telecommunication expenditures and to provide for charging back, if that is corporate policy.

Until very recently the SMDR and CDR capabilities that were available with CENTREX services lagged seriously behind similar systems and services that are widely used with digital PBX systems. This situation is now starting to improve, but there is still a good deal of dissatisfaction among telecom managers with the CDR support from some telephone operating companies.

The common approach to call accounting with digital CENTREX is for the telco to supply SMDR on magnetic tape from the central office system, on a monthly batch basis to the customer. The CENTREX customer, in turn, then sends the tape to be processed by a telecommunication service bureau, which prints monthly reports for distribution within the customer's organization. Although this procedure is better than no call accounting at all (which was essentially the situation with older CENTREX systems), it still falls far short of the needs of a well-managed, telecom-conscious organization.

A call-accounting subsystem should enable the CENTREX system administrator to make on-line inquiries (e.g., CDR for a 24-hour period or for a specific department); to define custom-tailored reports; to set the start and finish dates of a reporting period to coincide with the telco's actual long-distance billing dates; and to obtain a wide range of traffic data. In this way the SMDR output can be used to optimize the user's network, in such areas as WATS, 800 lines, and leased circuits, as well as to perform cost accounting.

The 5ESS has a separate applications processor (AP) that provides message detail recording to customers' locations. The MDR data are transmitted in real time from the main 5ESS processor to the AP over a packet-switched link. The AP stores these records and can process them to produce reports on all originating calls going over any type of private outgoing trunk, such as ETN, CCSA, WATS, or FX. Incoming call reports give details of all tandem-type calls coming in from other nodes on the customer's network. Public MDR reports are produced for all calls from the CENTREX system into the public switched network. Since the AP is on the telco's premises, it is unlikely that the user has any control over the format or content of call accounting reports produced by the 5ESS.

Northern Telecom has not yet delivered any improvement to the features of its Customer System Management, which provides heavily preprocessed data tapes from the DMS-100 system. The Dynamic Network Controller has been announced by Northern Telecom as a system to deliver on-line call detail records to customers' premises but this is not yet on general commercial release.

A few independent suppliers are taking advantage of the tardiness of the major CENTREX manufacturers and are selling powerful CDR systems to be attached to a CENTREX system, whether based on a 5ESS or DMS-100.

Telco Research has been selling its TRU system for CDR for over three years to CENTREX users who are on analog No. 1 and 1A ESS. It now has several options for digital CENTREX users. Customers served by Meridian Digital CENTREX may use either a synchronous or asynchronous data link between the DMS-100 in the central office and an IBM PC-AT or PS/2 at the user's location. These alternatives are both illustrated in Figure 3.1.

The synchronous option requires a Billing Media Controller (BMC), which is supplied by the Cook Electric subsidiary of Northern Telecom. The BMC emulates a tape drive and uses bisynchronous protocol over a data link, at 2.4 or 4.8 kb/s, for polling from the TRU system in the system administrator's office.

An alternative is to use the DNC, which is a new Northern Telecom system that can supply real-time SMDR or CDR data to multiple CENTREX customers. In this case, asynchronous data communication is used, up to 9.6 kb/s, and call detail records are stored and processed in the TRU system.

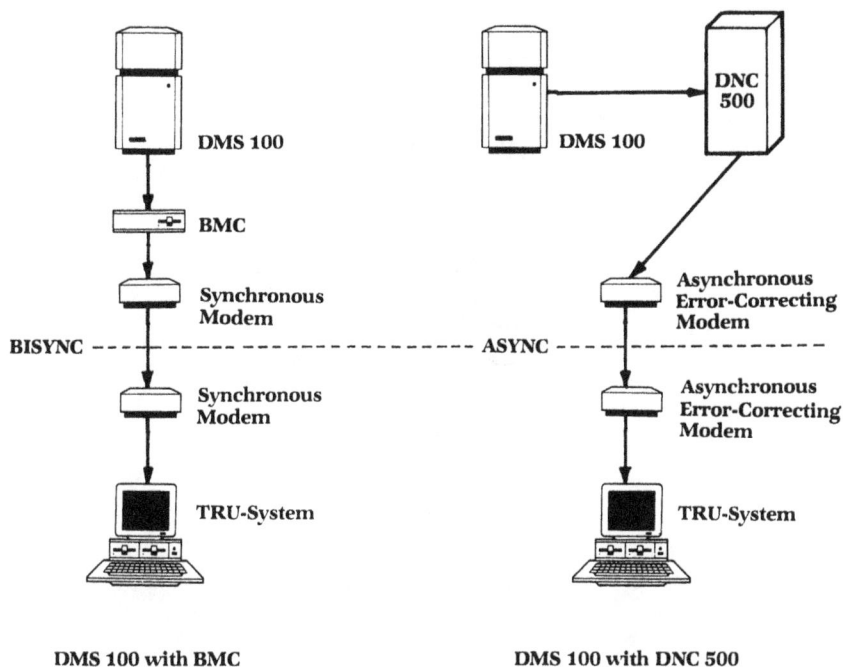

DMS 100 with BMC   DMS 100 with DNC 500

Figure 3.1 TRU System with Meridian Digital CENTREX.

AT&T's 5ESS CENTREX system provides real-time SMDR data from an AT&T 3B minicomputer. With this system an asynchronous link with error-correcting modems is also required. The 3B processor has the advantage that it can buffer data in the case of transmission failure. When the data link is restored then the 3B resumes the transmission of call detail data to the TRU system at the customer's premises. This configuration is shown in Figure 3.2.

Another example is the M3000 CENTREX Telephone Cost Management System (TCMS), which is supplied by Moscom Corporation. So far this software has been used to process CDR from the AT&T 5ESS. The TCMS program and data base resides in a personal computer at the customer's site and communicates with the 3B2 computer at the CENTREX serving office over a dial-up line. TCMS also incorporates an automated directory and a message center capability. The cost of this software ranges from two to six thousand dollars, depending on the number of CENTREX lines that are supported. Call records may be retrieved on an on-line basis, as well as both summarized and detailed call-accounting records.

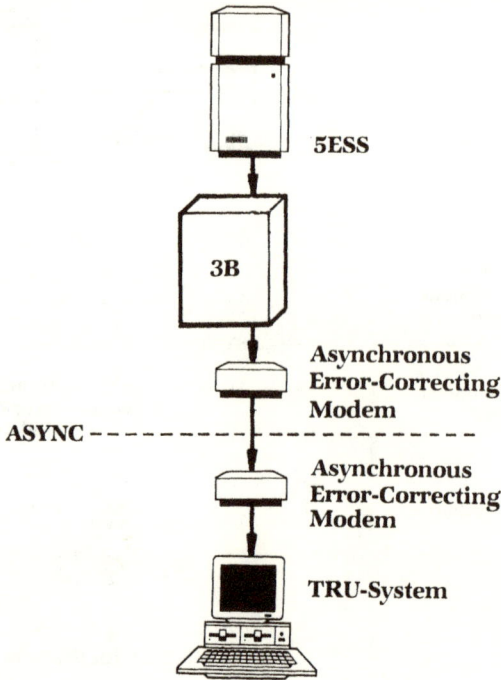

Figure 3.2 No. 5ESS with 3B computer for SMDR and CDR.

3.4 NETWORK MANAGEMENT

Network management is concerned with the minute-by-minute operation of a CENTREX system or network; the collection and classification of error messages (diagnostics); the testing of circuits and the restoral of service after a breakdown; and the collection of traffic data in order to optimize the network and plan for the future. For these applications, as with system management and system administration, the trend is very much toward putting these tools directly into the user's hands, although the great majority of CENTREX customers have not yet moved very far toward in-house network management.

AT&T has designed and implemented complete Switching Control Center Systems for a few large CENTREX users. One example is described briefly in the case study report about the state of Wisconsin, in Chapter 9.

Northern Telecom has announced the DNC-100 (known as the NM-1 in the SL-1 PBX environment) for network management from the customer's site. This system provides real-time network monitoring, with problem isolation, the management of problem resolution, and performance reporting. Network analysis and network design programs are becoming available for use with this system. The DNC-100 is intended to produce automated trouble tickets, keep a trouble ticket history file, and provide for customer-defined exception reporting.

Network Management is now recognized as a vital aspect of any major telecommunication system, but the provision of this capability to the users of CENTREX is still in its infancy.

AT&T has announced the concepts of its Unified Network Management Architecture (UNMA), but has yet to deliver any software to implement this plan. An increasing number of high-volume users are taking advantage of software defined networks (SDN), in which the availability of such services as WATS and 800 lines can be adjusted dynamically to accommodate varying traffic loads. The SDN can be implemented in conjuction with CENTREX.

It is likely that the Netview architecture, designed and being implemented by IBM, will be the dominant approach to integrated voice-data network management. The integration of Netview with central office systems and so with CENTREX service is likely to start soon, probably with switches from Siemens or Ericsson, because IBM has signed cooperative agreements with these two major telecommunication system manufacturers.

Chapter 4
Data Communication through CENTREX

4.1 CENTREX-BASED LANs

Most of the CENTREX vendors are now emphasizing the data communication capability of these systems. The concept of a central-office-based local area network (CO-LAN) is being sold quite heavily.

Data transmission through a CENTREX system requires the appropriate data interfaces, at the terminal, personal computer, or host computer (e.g., RS232 or RS422) and is subject to some limitations on cable length between the data device and the serving central office. On the gauges of twisted-pair wiring that are normally used within and between buildings, data transmission is generally possible without repeater equipment for up to 4000 feet on either side of the switch (i.e., 1.5 mi or over 2 km in aggregate). We must emphasize that this distance is not 1.5 miles on the map, but includes vertical and horizontal distances within buildings. In some cases, distances of up to three miles from the central office, or a remote concentrator, are feasible for data communications through digital CENTREX. The wiring in the building must be in good condition, with no bridge taps and with the minimum of gauge changes to ensure a good quality of data transmission.

The costs of providing data communication through CENTREX are, however, considerably higher per workstation than the corresponding charges for telephone service. There are often three reasons for the significant difference. First, the cost of the data interface module has to be factored into the charge for a data line. Such a module typically costs $300 to purchase and so probably adds at least $25 per month to the rental. Second, some types of data interface require two pairs of wires to the central office and are therefore rated at the equivalent cost of two voice circuits. Third, the telephone operating companies cannot afford to compete too aggressively with CENTREX data services against their own preexisting local data networks, especially data packet-switching services.

The use of pre-ISDN CENTREX to provide a LAN within a single building, or, as is even more common, on one floor of an office building, seems highly unlikely to develop as a popular application. Concerns about data security, system reliability, and monthly costs are all significant objections to the widespread implementation of CO-LANs.

Many analysts and designers concerned with LANs believe that some users will require occasional file transfers over a LAN, involving hundreds or thousands of kilobytes of data. For this purpose the 56 kb/s bit rate through the central office is not nearly fast enough. A LAN with a minimum throughput of 1 Mb/s is much more appropriate. Such a LAN may well use the twisted-pair wiring that is shared with CENTREX, but is a physically separate system.

4.2 METROPOLITAN AREA NETWORKS

It is possible to provide a citywide CENTREX data service through data transmission from one location in a city to another (through the nearest central offices providing CENTREX). This means that CENTREX could provide both local and metropolitan area networks for the use of an organization. The data interface devices used with analog telephones on digital CENTREX are essentially identical with those sold by the same manufacturers in conjunction with their PBX systems. In practice, asynchronous data speeds of up to 19.2 kb/s and synchronous data rates up to 56 kb/s are now being offered on digital CENTREX in many cities.

The use of the cost-effective network provided by a citywide CENTREX service as a MAN will usually be much more attractive to many organizations than a CENTREX-based LAN. This is brought out clearly in the case studies that are described in Chapters 6 and 8. A CENTREX system can provide bit rates that are comparable with those expected from public networks (i.e., up to 56 kb/s) and so does not present any bandwidth disadvantages. This is in marked contrast to the LAN situation, where a CENTREX solution is competing with other systems that can commonly offer data speeds of 4 and 10 Mb/s and can promise even higher bit rates in the near future.

With some telcos the approximate additional cost of a citywide CENTREX network is only one dollar more per month for each line than the cost of a local, one-location system. In these cases the use of that network for data communication is likely to be much less costly than comparable private (leased) analog or digital lines and may even be competitive with T1/DS-1 24-channel digital links.

4.3 CO-LAN SYSTEMS

Northern Telecom is the only one of the five present suppliers of CENTREX systems that is giving great emphasis to data communication through CENTREX in the pre-ISDN environment. AT&T and Siemens now present data communication solely in the context of ISDN, even though that capability is not yet widely deployed.

Northern Telecom provides a variety of data communication options through its DMS-100 switch (or the similar SL-100 PBX) by means of Datapath, which is a digital circuit-switched system. At mid-1988 there were 3500 data interfaces in use on Meridian Digital CENTREX in Canada, a number representing just over 0.5% of the CENTREX lines in service. This wide range of choice is illustrated in Figure 4.1.

Datapath uses only three types of line cards in the DMS-100. The digital line card (DLC) provides a range of up to 5.4 km (18,000 or 3.5 mi) over one pair of wires, based on time compression modulation (TCM), which uses a bit rate of 160 kb/s over the digital link to the central office. The DLC can now support synchronous communication up to 56 kb/s, which will be increased to 64 kb/s under the "clear channel" signaling conditions of ISDN. The asynchronous interface line card (AILC) provides an RS-422 interface, operable to 1.2 km, or 0.75 mi, over four wires, up to 19.2 kb/s. Where an interface to the analog side of a modem is required in the central office (CO) switch, this is provided by the standard analog line card (LC) that is used to support regular analog telephones. It should be noted that coaxial cable elimination is possible through the terminal interface (TIF) and control unit interface (CUIF). This means that small numbers of 3270-type terminals may be situated remotely from the cluster controller.

The operating range of Datapath can be extended in a number of ways. With a single CENTREX system the standard Datapath interface cards may be plugged into any one of the four remote modules that can be associated with a DMS-100. This can extend the range of one system to a radius of 100 mi (160 km) over a T1/DS-1 link. One DMS-100 central office can be linked to other central offices or remote PBXs using a network of leased lines (MSN or ETN), or over the public switched network, to create a wide area network (WAN) for data communication. Northern Telecom has demonstrated that Datapath can interwork with AT&T's circuit-switched digital service, based on the No. 1A ESS.

A number of features to expedite data calling are available on Meridian Digital CENTREX. These include speed calling, ring again, automatic dialing, and hot line (sometimes known as an automatic line or a "nailed-up" connection).

AT&T is selling two LAN systems for data communication that can be integrated with a CENTREX system and use pairs within a universal twisted-pair cabling system. Starlan is a small scale LAN, operating at a nominal speed of 1 Mb/s, that is widely used for personal computer communication in the office environment. ISN is a packet-switching system that supports synchronous and asynchronous devices, with a throughput of over 16 Mb/s. When an ISN system is installed on the telco's premises it is known as Datakit, but provides the same capability of supporting up to 1,600 data devices on one switch. Multiple ISN (Datakit) switches may be interconnected, by optical fiber, to form a data communication network.

We should note that it is possible to provide data communication through the older CENTREX systems by using modems, operating within the standard voice frequency range, or by using data over voice (DOV) units, which transpose

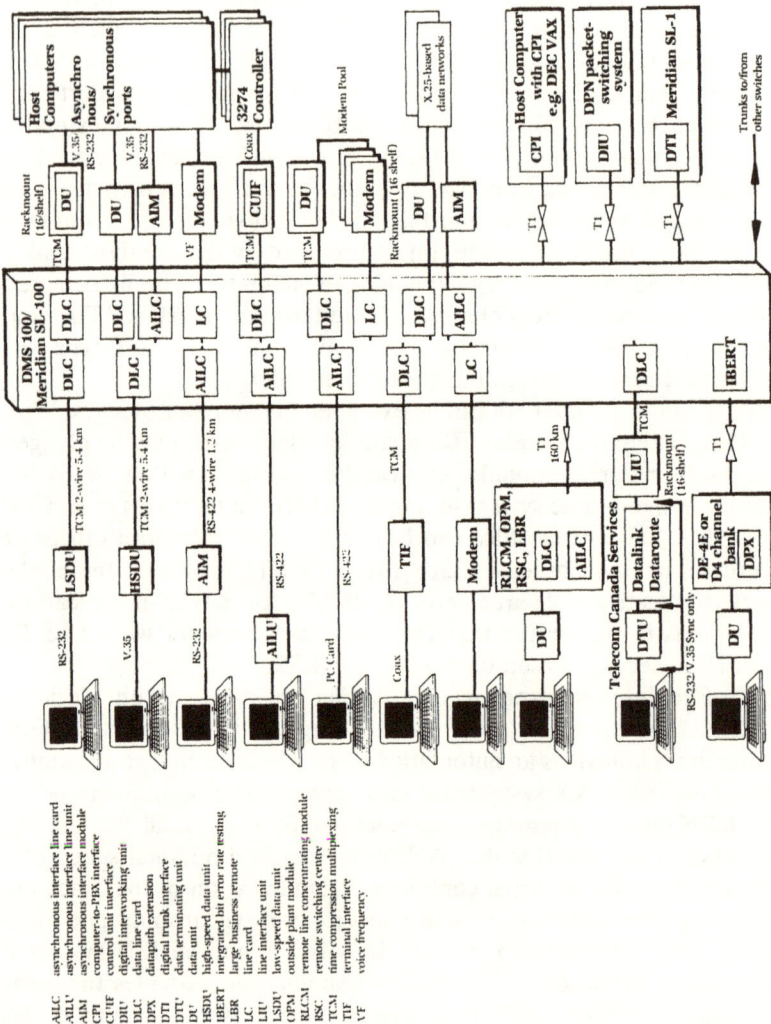

AILC   asynchronous interface line card
AILU   asynchronous interface line unit
AIM   asynchronous interface module
CPI   computer-to-PBX interface
CUIF   control unit interface
DIU   digital interworking unit
DLC   data line card
DPX   datapath extension
DTI   digital trunk interface
DTU   data terminating unit
DU   data unit
HSDU   high-speed data unit
IBERT   integrated bit error rate testing
LBR   large business remote
LC   line card
LIU   line interface unit
LSDU   low-speed data unit
OPM   outside plant module
RLCM   remote line concentrating module
RSC   remote switching centre
TCM   time compression multiplexing
TIF   terminal interface
VF   voice frequency

Datapath uses only 2 types of data line cards in the switch – the data line card (DLC) and the asynchronous interface line card (AILC) – to support a variety of customer premises equipment. Datapath also provides DMS connectivity to private branch exchanges, packet-switching systems, remotes, and host computers.

Figure 4.1 Datapath configuration on DMS-100.

the data signal into a frequency band higher than the 4 kHz voice spectrum. A few large users (e.g., Control Data in Minneapolis) have reported success with CO-LANs based on low-cost DOV modules, through CENTREX that is provided from crossbar (electromechanical) central offices. This may be viewed only as an interim solution that is becoming less feasible as the proportion of digital CO systems increases. In general we do not expect any significant use of CO-LANs until ISDN becomes widely available and unless the tariffs for transmitting data through the central office become more attractive than is generally true at the moment.

Chapter 5
CENTREX and ISDN

5.1 LINKING CENTREX AND ISDN

All of the major manufacturers of central office equipment are now trying very hard to suggest that CENTREX and the integrated services digital network (ISDN) are almost one and the same service.

At recent industry exhibitions, AT&T has described CENTREX-ISDN as "a challenging new telecommunication product" in various demonstrations and brochures. All of the literature from Siemens refers to its CENTREX offering on the EWSD switch as "ISDN CENTREX." Northern Telecom, in turn, has included the term "Meridian ISDN CENTREX" in at least one of its documents, instead of the more common Meridian Digital CENTREX.

Advertisements and presentations for CENTREX in North America have recently been emphasizing that users of CENTREX will have easy and early access to the full range of services promised by the integrated services digital network. All of the first seven field trials of ISDN with the Bell operating companies in the US were based on CENTREX-providing central offices. At this early stage it would appear that being a CENTREX customer may remove concerns about the costs and complexities of access, or nonaccess, to ISDN.

In an ideal world the question of ISDN access should not really be a primary criterion in making a decision for or against ISDN. The major PBX suppliers will be forced by fierce competitive pressures to provide ISDN interfaces, both hardware and software, as soon as the standards are well defined and the services become available. However, there are suggestions that ISDN may be available initially in some areas only to CENTREX customers, because of a marketing decision by the local Bell operating company. This might mean that some CENTREX users have access to the benefits of ISDN a year or two earlier than those who rely on an in-house PBX.

5.2 EVOLUTION OF ISDN

The official CCITT definition of an ISDN is "a network evolved from the telephony integrated digital network that provides end-to-end digital communications to support a wide range of services, including voice and nonvoice services,

---

to which users have access by a limited set of standard multipurpose customer interfaces." Two major emphases are that the ISDN concept is evolutionary, meaning that more developments are yet to be announced, and that the user-network interfaces are going to be standardized. It is fair to state that ISDN is now moving from theory to limited reality.

National and regional ISDNs will be interconnected to form a worldwide, public, all-digital telecommunication network to deliver voice, data, and other communication services in a standardized way. ISDNs are being implemented on the basis of digital transmission systems that have been coming into service over the past 20 years, digital switching systems that have been available since the late 1970s, and digital accesses that are now being developed.

5.3 ACCESS TO ISDN

A variety of "digital pipes," supporting different maximum bit rates, will become available to the users of an ISDN. Even though a pipe may have a fixed capacity, the digital traffic on that pipe may consist of a variable mixture of services and these will give the users access to circuit-switched and packet-switched networks, on a dynamically allocated basis. If a user is instantaneously employing less than the capacity of a digital pipe, the charging may be according to the aggregate bit rate actually used, rather than for the connect time.

The basic building block for an ISDN is the 64 kb/s B (bearer) channel, based on the present-day common standard for pulse code modulated telephony signals. Within an ISDN all signaling information for two or more communication channels will be carried on a separate D (data) channel.

Various numbers of B channels may be combined to form higher-speed digital pipes and these will be classified as H (high-speed) channels. Two H channels have been defined by the CCITT for use in North America. H0 combines six B channels to have a bit rate of 384 kb/s and H11 uses the full bandwidth of 24 B channels to provide a throughput of 1.536 Mb/s. For example, one H0 channel should be able to carry a color TV channel of teleconference quality, if signal compression techniques are used.

The B and H channels carrying customers' information are being described as "clear" channels, because no signaling data are encoded in those channels. The common analog voice-frequency telephone channel is now known as an A channel in this alphabetical hierarchy.

The "basic access" for ISDN is defined by the CCITT as two B (64 kb/s) channels and one D (16 kb/s) channel, for a total effective data rate of 144 kb/s. Basic access will be the universal interface for residential customers and small business locations.

For medium to large organizations the "primary access" will be the most common way into an ISDN. In North America and Japan primary access will provide 23 B channels and one D channel, with this primary version of the D channel carrying up to 64 kb/s, and one bit in each 193-bit word being used for synchronization purposes, for a total of 1.544 Mb/s. Later on, higher speed inter-faces, at such speeds as 45 and 135 Mb/s, will be required to provide higher capacity links, but these broadband ISDN standards have not yet been fully defined.

5.4 ISDN REFERENCE CONFIGURATION

One of the most important areas of ISDN standards has to do with the functional groupings and interface points at the customer's premises and in the neighboring network. This "reference configuration" is illustrated in Figure 5.1 as simply as possible.

The functional groupings in that diagram can be best described as follows:

TE1 = terminal equipment (e.g., digital telephones or integrated voice-data workstations) that complies with the ISDN user-network interface. In some documents the abbreviation ST (for subscriber terminal) is being used instead of TE.

TE2 = terminal equipment equipped with a pre-ISDN interface, such as RS232 or RS422.

TA = terminal adapter needed to connect non-ISDN devices (i.e., TE2) to the network.

NT2 = network termination on the customer's premises between the T and S interfaces side of the connection to the network. NT2 includes functions equiv-alent to layers 2 and 3 of the open systems interconnection (OSI) reference model and could well be a digital PBX, a LAN, or a terminal cluster controller. Both the S and T interfaces are based on two pairs of wires (i.e., four wires), one pair to send and one pair to receive. The U interface is a single pair (two-wire) interface.

NT1 = network termination at the customer's end of the digital local loop to the ISDN, performing functions at level 1 of the OSI model, including physical and electrical loop termination and signal conversion. The NT1 unit needs to have a different specification for each type of central office.

LT = loop termination at the carrier's end of the digital loop. This may be at a remote concentrator or a remote switching module.

ET = exchange termination is the central office (serving public exchange) com-plex, which may well be a combination of circuit- and packet-switching systems.

The various reference interface points—R, S, T, U, and V—may coincide with physical interfaces, but this will not always be the case. For example, in the US the boundary for customer premises equipment (CPE) typically extends further out into the network than is usual in Europe. The line of demarcation in the US

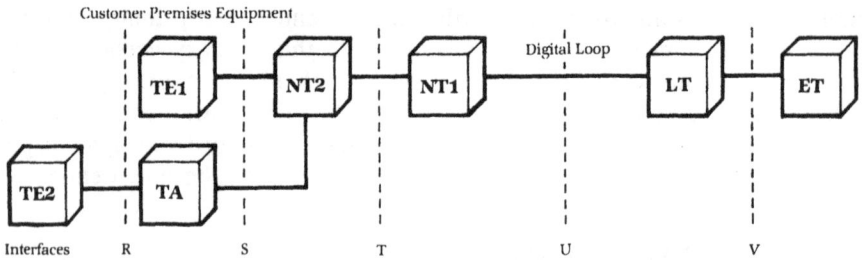

Figure 5.1 Network entities and interfaces for ISDN.

is at the U interface, which has been defined by the T1S1 committee, but is not specified by the CCITT. This means that the NT1 apparatus will be customer owned in the US. In Canada, at least for a while, the NT1 will only be available on rental from the telephone company, which will inevitably make ISDN services more expensive to the customer. In many cases the functions of NT1 and NT2 could be combined, typically in a PBX, performing as an NT12 termination. This combined system may also contain the TA protocol conversion to allow non-ISDN devices (TE2) to be connected to the ISDN.

5.5 ISDN SERVICES

For telephone and data services the ISDN should extend the enhanced features that are available with CENTREX and with modern PBXs—including call forwarding, speed calling (abbreviated dialing), conference calling, calling number and name identification, and real-time call cost information—across the complete network.

The new Group 4 digital facsimile machines can transmit a full page in about six seconds at 64 kb/s. Slow-scan video at that speed can also refresh the screen in only six seconds and electronic messaging systems can use the much higher data rates.

With the use of ADPCM, each B channel on ISDN can support one voice conversation and provide up to 32 kb/s for data, so the user can participate in simultaneous voice and data conference calls over one B channel. A full range of telemetry services, such as meter reading, remote alarms sensing, and environmental control can be carried on the D channel together with all the needed signaling data.

5.6 SIGNALING THROUGH ISDN

The use of out-band signaling right across the network is a fundamental feature of ISDN. This concept of common channel signaling at fast data rates leads to rapid call establishment, efficient use of bandwidth (providing the clear 64 kb/s channels), and the feasibility of the many services offered by ISDN. The ISDN-compatible system will have to process the link access protocol type D (LAP-D), for end-to-end signaling over the D channel (whether at 16 or 64 kb/s), and may also be connected to the common channel signaling No. 7 (CCS7) if that is extended beyond the carrier's nearest switch for customer-to-customer signaling purposes.

5.7 ISDN BENEFITS

For most telecommunication users in business there should be significant advantages in the ISDN, including cost savings, increased flexibility, and a wider range of services. The quality of data, video, and voice transmissions will be improved, with significantly shorter call setup times. In many cases the need for private lines or leased data circuits will disappear as the ISDN will effectively provide virtual private lines. The concept of software defined networks, where the bandwidth used and paid for by the customer is allocated dynamically, will be associated with ISDN.

An ISDN will enable the intelligent public networks and the intelligent customer premises equipment (such as digital PBXs and host computers) to function as one system, to an extent that has not been possible until now. This will allow businesses to take advantage of distributed networking intelligence, giving them the freedom to use the most economical solution at any point in time. The use of ISDN standards will lead to common service features among all the hybrid arrangements of networks and systems.

The simplest way to summarize the advantages of an ISDN is to state that all the features now available within a building from digital CENTREX or a digital PBX for the management of voice and data calls will be extended across the whole telecommunication network.

The long-term benefits of ISDN will necessitate that the business customer has easy access to the ISDNs of the various carriers and this access will be provided either through CENTREX or by an integrated PBX.

5.8 STATUS OF ISDN

Within the last two years the ISDN has moved from trials by manufacturers and within the carriers' own premises, to pilot services for commercial customers in several countries. Tariffs for ISDN service have been published by Deutsche Bundespost and a partial ISDN service is being provided commercially by British Telecom.

At least one Bell operating company had announced public ISDN tariffs (as compared with the customer-specific pricing which has already been established in a few cases) by the middle of 1988. AT&T is now advertising a full range of ISDN capabilities and plans to introduce ISDN service in a number of major urban areas throughout the US during 1989.

Southwestern Bell began to implement networks similar to ISDN for two major companies in the Houston area in the spring of 1988. Shell will have a 5500-station and Tenneco a 4000-station intelligent network, on ten-year contracts.

In the case of Tenneco, nearly 3000 of the terminals will be equipped with asynchronous terminal adaptors that will support data transmission over the D channel at 9.6 kb/s or over a B channel at 19.2 kb/s. These arrangements will facilitate PC-to-PC communication, access to a modem pool at the central office, and connection to IBM mainframes through ASCII-3270 protocol conversion.

This ISDN application at Tenneco will use packet switching to both IBM and DEC-VAX systems. Up to 128 data terminals, on D channels at 9.6 kb/s, can be served by one host connection at 64 kb/s on a single B channel.

IBM 3174 terminal cluster controllers will be connected through B channels, at 64 kb/s, to IBM 3725/45 communication controllers, using AT&T model 7500 digital telephones at each end. Remote graphics design workstations will use a similar arrangement.

Tenneco finds that the basic rate access speeds provided by ISDN are more than adequate to satisfy its current data communication needs. The exceptional requirements for bit rates higher than 64 kb/s will be met by specialized LANs that will have gateways to the ISDN. In this case digital CENTREX provided all the voice communication functionality that was needed for 5000 employees spread over Houston and the surrounding suburbs. ISDN appears as the icing on the cake that will cost-effectively provide most asynchronous and synchronous data communication solutions. C&P Telephone has signed a somewhat similar contract with the state of Virginia, except that the potential ISDN-related services are identified in a much more detailed manner in this case.

AT&T has published two case studies that show that CENTREX-ISDN can be more cost-effective than the purchase of a number of PBX systems. It is significant that both of these cases involved organizations that have personnel in a number of locations, so that CENTREX-ISDN would be used as an intelligent voice communication network. In addition, there would be extensive use of the network by data workstations, both 3270 terminals and personal computers.

There seems to be a general consensus that for those users who have proven that CENTREX is the right solution then the migration to ISDN will be even more advantageous. At the level of voice-only service there is frequently little choice between a PBX-based solution and the new CENTREX-ISDN. However when the increasing use of the network for PC communication and the prospect of using one ISDN channel for up to eight voice and data devices are factored into the calculation, then ISDN shows a cost advantage of 20% or more. Of equal importance to any potential cost savings are the improved user-oriented services and network management capabilities that will be available through an ISDN.

5.9   ISDN FIELD TRIALS

Two field trials with ISDN started in the US in late 1986. Mountain Bell is providing ISDN service from a Northern Telecom DMS-100 in Phoenix to several offices of the state government of Arizona. In the northwestern suburbs of Chicago, Illinois Bell started the best-known ISDN trial with 50 basic rate lines from an AT&T 5ESS switch to the head office of McDonald's Corporation. This field trial has now progressed to a commercially tariffed service with hundreds of ISDN lines to multiple locations.

All seven of the Bell operating companies and several of the GTE companies had at least one ISDN trial underway by the end of 1987. Bell Canada also started an ISDN trial in late 1987 with three departments of the federal government in Ottawa.

Some of these ISDN pilot trials, which are typically intended to run for about 12 months each, may be best described as technology trials while others are primarily application trials. All of the trials are based upon central office systems from the three major suppliers, namely AT&T, GTE, and Northern Telecom, and nearly all are providing effectively enhanced CENTREX services. The majority of the cost and convenience advantages that have been identified so far relate to data communication, partly because these are all very small islands of ISDN within a sea of analog telecommunication services. Some of the significant advantages that ISDN will bring to voice and facsimile communication will only become apparent as wider integrated digital networks are implemented. A few details that have been published about two of the twenty or more ISDN trials in North America should help to confirm the potential for this major development in CENTREX service.

5.9.1   ISDN Trial by Pacific Northwest Bell

Pacific Northwest Bell conducted an ISDN trial with the US National Bank of Oregon (USNBO) in Portland. This trial is based on a DMS-100 switch, which supports nearly 8000 non-ISDN customers and provides 200 basic rate access lines for the bank. These lines go to the bank's head office and to thirteen branch locations within the city.

One application involves the connection of workstations to an IBM 3270 system, using either the D or B channel through the local ISDN. This example of SNA/3270 connectivity is illustrated in Figure 5.2. In this case a terminal adapter board in each PC emulates a 3274 cluster controller. An X.25 packet handler is associated with the DMS-100 and this concentrates the data traffic over a 9.6 kb/s link to the 3725 front-end processor.

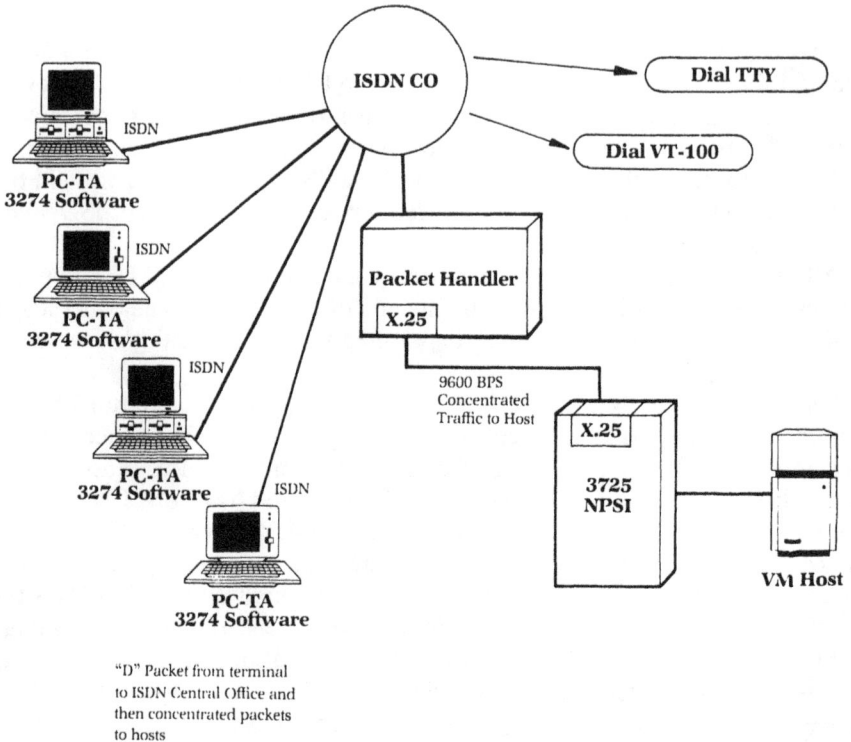

Figure 5.2 SNA/3270 connections through local ISDN.

An extensive number of local area network arrangements were tested with USNBO, as illustrated by the LAN elimination scheme in Figure 5.3. Here the personal computers are linked to the central office over D channels and a common B channel handles traffic to a central file server.

The feedback from the users of this arrangement shows that it works well for spreadsheet and word-processing applications and so meets the needs of over 90% of PC users. However, the speed restrictions of basic ISDN, particularly the

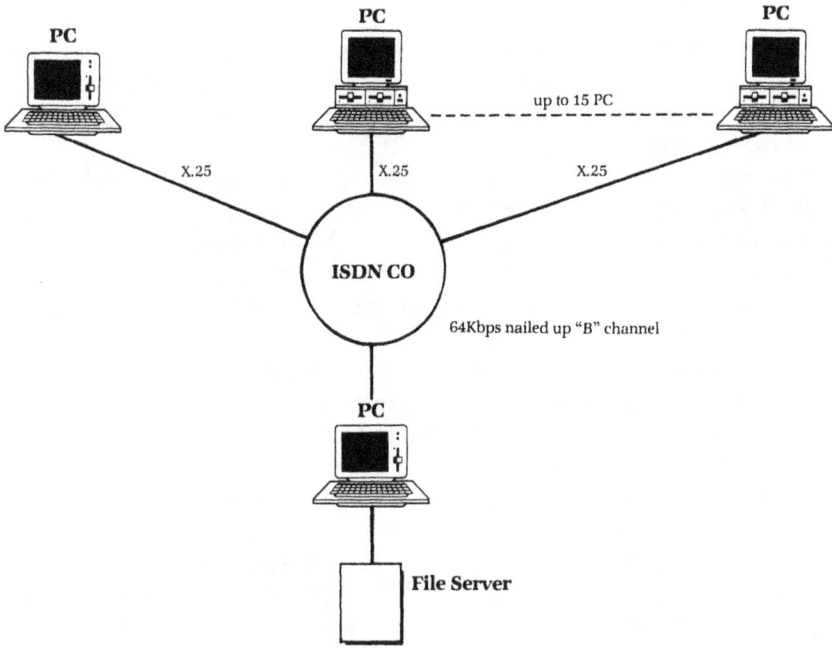

Figure 5.3 CO-LAN over D and B channels.

16 kb/s limitation over the basic D channel, preclude effective use of significant file transfers. B channels were extensively used in this Pacific Northwestern Bell field trial for PC to LAN, LAN to LAN, and LAN to host computer connections.

For applications that involve the connection of asynchronous workstations to a remote time-sharing service the bottleneck associated with 1200 or 2400 b/s dial-up modems has been eliminated by using packet-switched data over shared D channels.

5.9.2 ISDN Trial by Northern Telecom

The first ISDN trial to combine a PBX and CENTREX was announced by Northern Telecom in June 1988, in cooperation with Glaxo Laboratories in Durham, NC. An SL-1NT PBX in one office location is linked by a primary rate access to a DMS-100 operated by the local GTE company. A second Glaxo office has a number of basic rate accesses through Meridian Digital CENTREX to the same central office. In this way voice and data communication will be integrated between the two locations.

5.9.3 ISDN Trial by Bell Canada

The trial of ISDN by Bell Canada, in Ottawa, provides a variety of services to 84 users within seven different groups from four government departments. This trial is based on one of the four DMS-100 systems that support 90,000 users in the national capital region in a digital CENTREX network.

Each user in the Bell Canada trial has a personal computer and a Northern Telecom T 2317 digital telephone as an integrated workstation. Two users typically share one ISDN line, with the telephones on separate B channels and the two PCs sharing one D channel. Three valuable applications that use the features of ISDN have emerged in these government departments, involving call manager, screen sharing, and wide area networking.

Call Manager is a software package (developed by Bell-Northern Research), that runs on a PC to provide an on-line directory to permit auto-dialing from the keyboard, to log incoming calls, and to facilitate the calling-line-indication feature. The receptionist for a department with a high rate of incoming calls can now have information about who in the organization should field which call appear quickly on a screen and then transfer the call correctly with a single key stroke.

Screen sharing is an aid to users of a large LAN. A non-ISDN user on the LAN who is having a problem can now call a technical analyst, set up a data link through a gateway to the ISDN from the LAN, and demonstrate the problem directly on the analyst's screen. This feature greatly improves the productivity of both the resource person and the network users.

For wide area networks between buildings in Ottawa the ISDN is an order of magnitude faster than the use of a modem on the public switched network. For most of the users of a personal computer, the ISDN replaces up to five programs and different interface cards with just one program and a terminal adapter board, which is evidence of true integration.

The results from these trials, and others that have been undertaken over the last two years, suggest strongly that if ISDN can be provided with the basic rate access costing less than twice a CENTREX line, then CENTREX-ISDN will prove to be attractive for many business users.

Chapter 6
Evaluating CENTREX Systems

6.1 FINANCIAL ANALYSIS

A full financial analysis to compare the life-cycle costs of a PBX, or a network of PBXs, with the costs of providing the same communication facilities through CENTREX services must be specific to a given organization and could take several weeks to complete, if it involves several thousand users and a number of different locations. There are some important common factors in such analyses that we can summarize here.

The comparative analysis should cover a period from five to ten years into the future. It is now fairly common to assume a system life of seven years for a modern, digital PBX system. Some government organizations seem to prefer the assumption of a ten-year life, while quickly changing businesses should probably estimate for a system life of only five years.

The total life-cycle costs of a PBX system can easily exceed twice the initial purchase cost over a seven-year period. In general there will be higher costs with a PBX than with CENTREX for in-house personnel to manage and maintain the system, for electrical power and floor space, and for access to public networks. In addition the cost of a maintenance contract on a PBX and all of its attached apparatus is often at the level of 5% of capital cost per year.

A CENTREX solution may be attractive to an organization that expects to make a major office move to another building within a few years or that is expecting greater than average growth (or contraction) rates in its office population. On the other hand, in a fairly stable situation, the fixed, one-time, capital cost of a PBX, purchased in a very competitive market situation at around $500 per local line, may be preferable to the ongoing rental cost of a CENTREX system.

Another advantage for an organization that uses CENTREX from several locations in one city will be the possibility of low-cost access to long-distance services. There is a definite trend for the cost of local leased lines and services to increase more rapidly than long-distance costs. This means that it may become increasingly expensive to collect long-distance traffic from several locations into one main PBX for access to the competitive, long-distance networks. CENTREX may well become a cost-effective, citywide collection agency for voice, data, and image traffic and provide gateways to competitive public switched services and to a private integrated digital network.

These considerations are illustrated, together with other financial and operational factors, in the following two case studies.

6.2 MULTILOCATION CASE STUDY

A regional government authority, which covers the eastern suburbs of a large urban area, is responsible for a population of about 420,000 people. This regional government has some 6400 employees (many of whom are part-time) and of these approximately 2500 are classified as managerial, professional, and office personnel. The municipality has five major office complexes (see Table 6.1).

Table 6.1

Location	Number of Phones	Map Code*
Civic Center	900	C
Social Services & Data Center	200	D
Engineering & Transit Garage	300	E
Finance Department	400	F
Personnel and Training	200	P

*See map in Figure 6.1.

Several other, smaller offices had their own switchboards or key telephone systems with tie lines to the nearest main office. These locations include a number of libraries. Off-premises extensions (OPX) were used to other locations such as smaller garages and sports centers. Each of the main locations had an out-of-date analog electronic PBX with groups of analog tie lines to the Civic Center. The network configuration is illustrated in Figure 6.1.

A total of 2200 telephones were served by the switched network and this number is expected to remain stable over the coming years. All the incoming calls to the local government offices come in from the nearest central office to the Civic Center location, where there are presently eight central attendant positions serving 94 incoming and 52 outgoing trunks.

We should note that this widespread suburban area is served by five different central offices (wire centers) and that each of the four other main office sites has its own outgoing only CO trunks, as indicated on the diagram. By 1987, telephone

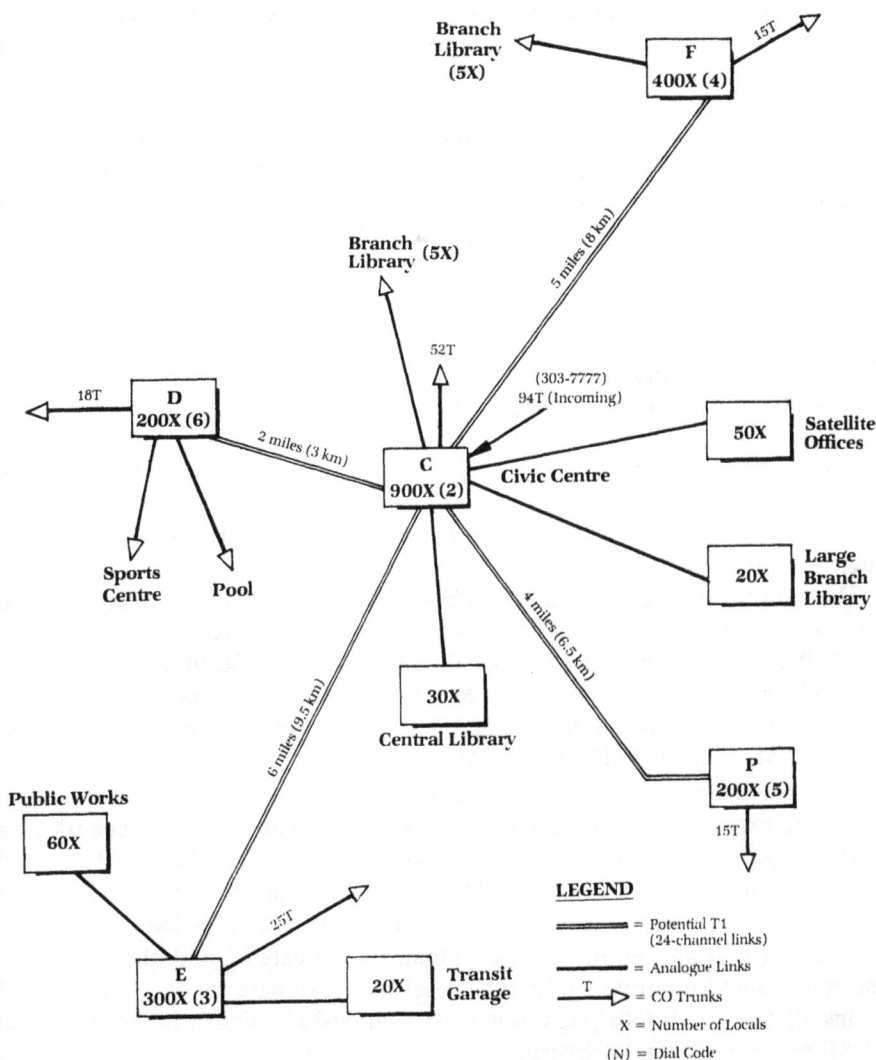

Figure 6.1 Network configuration for suburban government.

traffic was estimated at just over one million incoming calls each year, giving a peak load of nearly 110 calls per operator-hour.

The analog network, based on aging PBXs, was obviously overloaded and providing poor service to the public. The telecommunication manager was able to show, in 1986, that nearly 10% of all incoming calls were not completed, due to lack of capacity on the circuits between the main Civic Center switchboard and the other network nodes. Approximately another 10% of calls were thought to be not getting through the central office to the switchboard attendants, due to all of the incoming trunks being busy at certain times. By mid-1986, the telephone users were also complaining frequently about the quality of voice communication over the network, probably because there could be three analog lines and switches in tandem between a small office and the serving central office on an incoming call.

6.2.1 Cost Analysis

A star network, based on digital PBX systems, and digital (T1) 24-channel links between main nodes was considered as one option. The costs of such a PBX-based network, assuming an outright purchase of all the PBXs and EKTS, together with new cabling in all office locations, are summarized in Table 6.2(B). This PBX network would have a three-level structure, in that some of the smaller PBXs and some EKTS would be linked to a main node and this, in turn, would be linked through a multichannel digital line to the Civic Center hub.

The estimated costs for a digital CENTREX alternative are given in Table 6.2(A), where the monthly rental for each line includes the cost of a systemwide network, based on the five serving central offices. One significant point is that the CENTREX-based solution assumes the employment of only two central attendants in the busy hour, because all of the CENTREX lines would have DID, as well as direct outward dialing (DOD), capability.

The cost estimates matrix of Table 6.2 gives a fairly detailed comparison between the expected costs for a digital CENTREX solution and a network of a number of digital PBX and hybrid-key systems (e.g., the IBM 9750 and Redwood Business Communications Systems). The PBX-based alternative would include four T1 (24-channel) links between the major locations and some leased lines. We have assumed that all in-building telecommunication cabling and all of the telephone sets would be purchased in both cases. The estimate for the IBM Type 3 cable installation (with four pairs of wire to each desk) is slightly higher than that for the three-pair CENTREX wiring.

We have also assumed that CENTREX rental costs will be held steady over a five-year contract by the telephone company and have brought these future rental expenditures to an approximate present value by assuming a discount rate of 10%. Most other costs (e.g., personnel) have been assumed to be steady, because inflation and other increases will probably balance out the discount rate. A personnel requirement of ten is reasonable for the PBX alternative (including eight attendants).

The CENTREX line rentals in this table are based on multiwire center tariffs, at $20 per month for each of the first 150 lines on each CO and $16 per month for

Table 6.2 Cost Analysis: Local Government Network
(all amounts in thousands of dollars)

A. CENTREX Costs	Year 1 ($k)	Year 2 ($k)	Year 3 ($k)	Year 4 ($k)	Year 5 ($k)
CENTREX line rentals	460	418	380	346	315
Cabling installation	375				
1800 Single-line phones	180				
300 Feature phones	75				
100 Executive phones	45				
Phone set maintenance	21	21	21	21	21
Data communication lines	9	11	13	15	17
Personnel (4)	120	120	120	120	120
Attendant consoles	8	8	8	8	8
ASR and SMDR charges	24	24	24	24	24
Enhanced feature package	6	6	6	6	6
Initial setup charge	20	2	2	2	2
Annual Totals	$1343	610	574	542	513

CENTREX Total for 5 Years					$3582k

B. PBX Costs	Year 1 ($k)	Year 2 ($k)	Year 3 ($k)	Year 4 ($k)	Year 5 ($k)
PBX purchase (including software)	1540				
Cabling installation	450				
1800 Single-line phones	180				
300 Feature phones	75				
100 Executive phones	45				
Data Com modules (buy)	6	3	3	3	3
System maintenance	113	113	113	113	113
CO trunks (220)	105	95	86	78	71
T1 links and leased lines	120	109	99	90	82
Personnel (10)	240	240	240	240	240
Annual Totals	$2874	560	541	542	509

PBX Total Cost for 5 Years					$5008k

each line over the base level of 150. Data communication lines through CENTREX
are calculated from a tariff of $60 per month per line starting at 12 lines in the
first year and growing to 24 data terminations in year five. The additional charges
on CENTREX for automatic route selection and station message detail recording

are 50 cents per phone per month for each feature. The enhanced feature package provides speed dialing lists and multiline conferencing capabilities.

6.2.2 Equipment

Three grades of telephone set would be used in this network. The majority of personnel in the offices would have a standard 12-key single-line DTMF set with a recall button. A total of 1800 of these telephones would be needed. Three hundred feature phones, with the hands-free (microphone-loudspeaker) feature and a minimum of four function keys would also be purchased. Senior officers of the municipality and their assistants would be provided with a more expensive telephone, having at least 16 single-level programmable keys, a 40-character alphanumeric display and a twenty-number memory. Approximately 100 of these expensive telephone sets would be needed for the system, for an overall total of 2200 sets, of which 82% are the simple single-line telephones.

Previously, several senior managers in these offices had a second telephone on their desks in order to provide a direct outside line. These extra telephones would be conveniently replaced by DID service to 32 offices in the Civic Center and up to 16 offices scattered across the four other main buildings.

The telecommunication manager intends to purchase and install a central CDR system as soon as possible. This system will do overnight SMDR data collection from all of the locations on this regional network.

6.2.3 Data Processing

The data processing center for this regional government is located about two miles to the west of the Civic Center but is served by a different wire center.

Data processing activities have grown tremendously in this borough since the department was established in 1969. A wide range of applications is now supported on two IBM-compatible NAS 7/72 systems, which are actually made by Hitachi. This computer is roughly comparable to an IBM 4381 system, with 16 MB of main memory and 10 GB of on-line disc storage. In all, 85 people are employed in the DP department.

Most of the borough's applications run as on-line systems, being written in COBOL, using CICS, with the VM/DOS/VSE operating systems. There are 350 visual display units, equivalent to the IBM 3278 terminal, distributed across the various council offices. Most of the VDUs have an attached thermal page printer. The majority of these 350 screens have been supplied by Telex and each one is connected by a direct coaxial cable to a terminal cluster controller in its building.

Some 180 IBM-compatible Personal Computers have been installed by the borough. One hundred of these have Irma 3270-emulator boards and these are also connected to a controller (like the 3274) and can be used as 3270 terminals. Each PC has a locally connected Epson printer.

Prior to the installation of a new digital network the terminal cluster controllers were connected to a front-end processor (from NCR-Comten), through a star network of point-to-point analog private leased lines. Multiplexers were used, where necessary, to support multiple controllers in the one building. Most of the controllers were then connected to the host computer at a data transmission speed of 2400 b/s.

6.2.4 Integrated Voice-Data Network

Two alternatives were considered to enable sharing of the digital links by voice and data channels. A number of integrated networks use drop-and-insert multiplexers for the data links, on the network side of the switches. The second alternative, that was considered seriously, is to route the data traffic through the network using data interface modules between the PBXs or CENTREX and the controllers and computers. In some cases this solution would require a multiplexer to enable two or more controllers to share one 56 kb/s channel. The 3274-equivalent controllers owned by the borough support a maximum transmission speed of 9.6 kb/s in remote mode. Separate tenders for the multiplexers needed for this second possible solution were requested from data communication equipment suppliers and these have not been included in our cost comparison (because we consider that the cost would be the same regardless of whether a CENTREX-based or a multiple-PBX solution is chosen).

This second option of using the digital switching system(s) as the interface to an integrated voice-data network proved to be less costly than buying drop-and-insert multiplexers. This decision led to the adoption of the data network shown in Figure 6.2, where the model 1262 multiplexers are supplied by General Datacom. All the terminal cluster controllers will have a 9.6 kb/s digital data link to the host computer, which will significantly improve the system's response time for all of the users. Where only one controller is attached to the network at a location there is no need for a multiplexer, because the data interface module (on CENTREX) or the digital telephone (on a digital PBX) can provide bit rate adaptation up to 56 kb/s.

The data links through CENTREX or the PBX systems are permanent, "nailed-up" connections (sometimes known as "hot-line" connections) and will not be seen as switchable by the users. Since there are relatively few permanent data links for any one office there is no risk of data connections overloading the capacity of the voice system in any way. Note that up to 128 VDUs can be supported by one 56 kb/s circuit through the network because of the concentration effect of the terminal controllers and the multiplexers used with the digital interfaces or telephones.

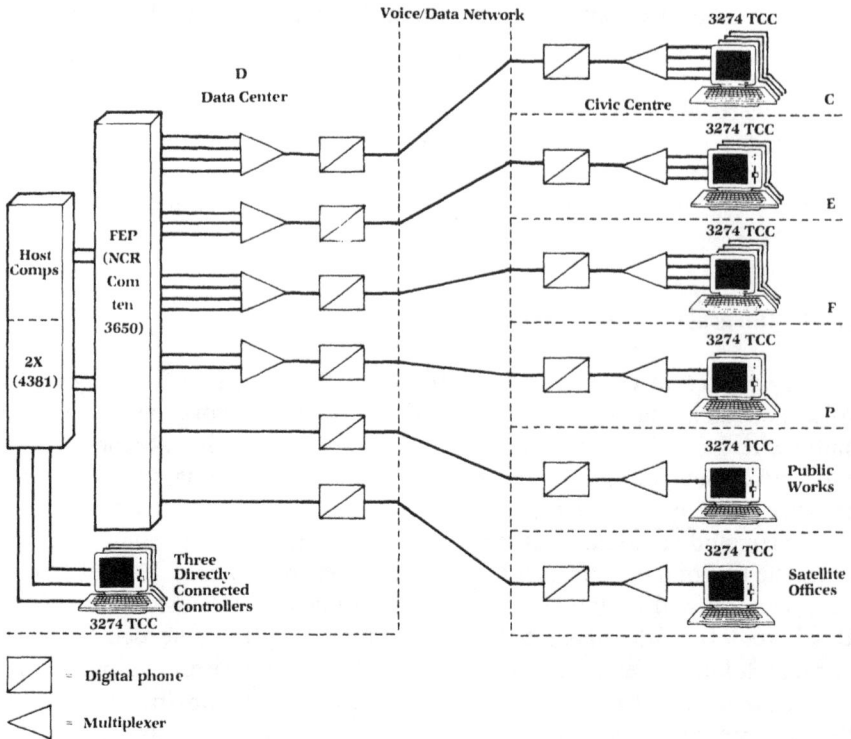

Figure 6.2 Data network for local government.

This data network is considered by the borough's telecommunication manager to be just the first stage of integrating voice and data from CENTREX. The borough has an automated library system, which is running on a GEAC computer at the data center, with analog data lines to all the libraries in the region. There is also a network of word processors that uses dial-up telephone line connections for intercommunication. In addition, as office automation applications are implemented, the total number of communicating data workstations could grow from 450 to about 900 units over a five-year period. The digital CENTREX network that has now been implemented has plenty of capacity to handle the anticipated growth of computer-based applications, particularly as voice traffic is expected to increase minimally in the future.

6.3 SINGLE-SITE CASE STUDY

This example is based on the head office of a medium-sized chemical company, with 400 telephones, which is experiencing a growth rate of 10% in personnel each year and has all of its office employees in one building. This company has no intention of linking terminals or personal computers to a switched network and is, therefore, comparing a voice-only PBX system to CENTREX for its single location. Table 6.3 is a cost analysis table.

Table 6.3 Cost Analysis: Head Office Location
(all amounts in thousands of dollars)

A. CENTREX COSTS	Year 1 ($k)	Year 2 ($k)	Year 3 ($k)	Year 4 ($k)	Year 5 ($k)
CENTREX lines	81	81	81	81	81
Cabling installation	50				
Single-line phones	30				
Feature phones	20	11	10	9	8
Executive phones	20				
Personnel (2)	60	60	60	60	60
Console	4	4	4	4	4
Additional features	12	12	12	12	12
Initial system setup	20				
Telephone maintenance	4	4	4	4	4
Annual Totals	$301	172	172	172	172

			CENTREX Five-Year Total		$989

B. PBX Costs	Year 1 ($k)	Year 2 ($k)	Year 3 ($k)	Year 4 ($k)	Year 5 ($k)
PBX purchase	240				
Cabling installation	50				
Single-line phones (300)	30				
Feature phones (60 initially)	20	11	10	9	8
Executive phones (40)	20				
System/phone maintenance	16	16	16	16	16
CO trunks (48)	23	23	23	23	23
Personnel (3)	80	80	80	80	80
Add-on line cards	0	5	5	5	5
Annual Totals	$479	135	134	133	132

			PBX Five-Year Total		$1013k

In this case we have estimated that a medium-sized PBX without data-handling features can be obtained at a purchase price of $600 per line (e.g., a Mitel SX-2000S or a Northern Telecom SL-1ST), compared to CENTREX at $18 per line for the first 160 lines and $16 per line for the other 240 lines at cutover. Because this customer will not have more than 1000 CENTREX lines the best possible tariff cannot be obtained. Most of the other costs are calculated on the same basis as in the previous case study.

We have assumed that all of the growth will be in the feature phones and have shown all of the telco tariffs at level costs, because the system growth will just about balance any discounting of future expenditures into net present value. In the case of the PBX we have allowed a further $5000 per year to provide the additional cards to support 40 more phones each year and have left the maintenance costs level, which means the actual expenditure is increasing by about 10% annually.

Based on these reasonable assumptions the aggregate five-year estimates for the PBX and CENTREX alternatives come to totals which are very similar, such that CENTREX would cost about 99% of the total for the voice-only PBX. It is interesting to note that if these calculations are extended to a system life of seven years (with no growth in the last two years), which is quite reasonable for a digital PBX system, then the net present costs come to $1.32 million for CENTREX compared with $1.26 million for the PBX.

We believe that most senior managers would choose the PBX option over CENTREX in this situation, primarily because there is still greater control over the corporate future in that case. However, we should emphasize that this second example shows how sensitive the CENTREX versus PBX decision is to a number of factors. If a 400-line PBX is quoted at $700 per line (before telephones and cabling) or if the telco can reduce the cost of each CENTREX line by 20% (perhaps by sharing one ISDN basic rate line between two phones, in the near future) then the decision would clearly swing in definite favor of CENTREX.

6.4 TECHNICAL CONSIDERATIONS

The major advantages and disadvantages of CENTREX services are described in Chapter 1. Some more concerns, with a slightly technical flavor, should be considered and are addressed in the following paragraphs.

Many PBXs are now advertised as "nonblocking" systems. This means that the PBX can be configured with sufficient hardware such that all the attached devices (whether telephones, data terminals, or computers) can be active simultaneously. Very few organizations really need a truly nonblocking switch and even with a nonblocking PBX there is no guarantee that a call attempt can be completed. An obvious condition that will prevent call completion is that the destination may be busy. Another important consideration is the busy hour call attempt capacity of the system, which depends largely on the power of the processor(s) used in the PBX. If the users of the system generate large numbers of short calls, then some users may not be able to obtain dial tone in the busy periods, due to the BHCA capacity being exceeded. Several PBX manufacturers, such as Mitel and Northern Telecom, have recently announced significant upgrades to the memory size and speed of their processors to handle this specific problem. In any case, it is at least possible to obtain a nonblocking PBX system as a form of long-term insurance against rapid growth of demand from telephone and data workstation users.

CENTREX systems are always implemented as blocking switches. Because the complexity of a switching matrix increases approximately on a square law basis proportionate to the number of attached lines, it is not economically feasible to install a nonblocking large central office. Some CENTREX implementations will, in fact, have concentration (i.e., fewer outputs than inputs) at two levels in the system. A local concentrator may be installed on or close to a customer's premises and the central office system, in the telco's equipment building, is designed on the assumption that only a fraction of its ports will be active at any point in time. Even though blocking is only part of the story, it is important to understand what the blocking probability is in the CENTREX system and how that relates to the traffic pattern of the prospective user organization.

It is also worth noting that the cabling run between two communicating devices on a CENTREX system will be much longer than the wiring distance with a PBX, which is commonly in the same building as all of its users. This consideration means that the CENTREX-based connection has a greater exposure to potential trouble than a totally in-house connection. In this case, trouble may mean a physical break in the line, perhaps due to outside construction work, or electrical interference with a stream of sensitive, high-speed data.

On the positive side for CENTREX we should remember that central offices are all equipped with high-capacity batteries and diesel-electric generators for power backup purposes. So for situations where telephone service is a vital concern, such as a hospital or a fire station, CENTREX may well provide greater reliability than is feasible with a customer-owned PBX.

6.5 CHECKLIST OF QUESTIONS

The following list of ten questions, which require a simple yes or no answer, should help you to decide whether your organization is a candidate for CENTREX service. This simple checklist will not, by any means, make the final decision regarding the choice between CENTREX or PBX, but your answers should point you in one direction or another.

1. Does your organization have two or more offices in one city (with over 100 phones in each office)?

2. Is it significantly easier to obtain an annual operating budget rather than a capital budget for the purchase of telecommunication equipment (i.e., does your company much prefer "to pay as you go")?
3. Is your senior management overly concerned about the cost of telecommunication personnel (analysts, technicians, or operators)?
4. Are you finding it especially difficult to recruit competent telecommunication personnel?
5. Do you wish to provide a centralized operator system to help your customers or employees?
6. Do your offices occupy especially expensive floor space?
7. Do you expect to move a major office within the next three years?
8. Is your company in a business where frequent changes in the number of personnel are expected?
9. Are you willing to pay the telephone operating company to configure the corporate telecommunication network and to administer all of your moves, additions, and changes?
10. Is your organization's management concerned about making a commitment to a PBX vendor that may not be able to keep up with rapid technological change?

Two or more Yes answers to the above questions indicate that your organization should at least consider digital CENTREX service when it becomes available in your location. If you have five Yes answers then you should start a serious evaluation of CENTREX versus a PBX-based solution.

Chapter 7
CENTREX in Financial Services

7.1 CENTREX IN BANKS

The larger companies in the dynamic financial services industry are obvious candidates for CENTREX services, because they have offices in many different locations and are now typified by a high rate of moves and changes. The various components of this industry may have a higher proportion of CENTREX telephones than any other industry in North America. Examination of a few case studies is appropriate for this chapter.

The largest commercial user of digital CENTREX service in Canada is the Royal Bank, with 11,000 lines in nine cities across the country. Six thousand of the Royal Bank's users are in the Toronto area. This bank pays an average of $25 per month for each CENTREX line, which is almost exactly half the cost of a single business telephone line. CENTREX III is used by the bank in major offices and also in bank branches in the main urban areas. For example, in Calgary, Alberta, the Royal Bank has 40 branches on CENTREX, as well as its regional office, because Alberta Government Telephones is flexible in terms of the minimum number of lines that may be connected to one CENTREX host switch. A few branches are more than 5 km (3 mi) away from the serving exchange, a distance which is beyond the range of the electronic business set, even with the use of loop extenders. In these cases electronic key telephones, such as Northern Telecom's Vantage system, are a viable alternative and work well with the direct inward dialing feature.

7.2 WELLS FARGO STAYED WITH CENTREX

Wells Fargo Bank has its head office in San Francisco and conducts banking business throughout California. The bank has some 18,000 employees, nearly 470 branches, and more than 1000 automatic teller machines. Statewide telephone service is provided by a private tandem network, using Northern Telecom's ESN software, that links 13 different CENTREX systems and nearly 60 PBXs. About one-third of the bank's telephones are presently provided by CENTREX services, some of which are based on switches made by AT&T and in all cases the telephones are directly connected to the serving CO, without any use of remote switching modules. Five SL-1XT PBX systems and one DMS-100, which is in San Francisco, form the nodes for this network.

Centralized attendant service is split between two locations (i.e., San Francisco and Los Angeles), with a maximum of only six attendants needed to assist the 18,000 telephone users during the busy hours. All requests for moves, additions, and changes from the branches and administrative offices are processed centrally by a team of five people in the telecommunication division of the Wells Fargo Bank and are then passed on to PacificBell (PacBell) for implementation. In a similar way, any problems with telephone service at the bank's offices are handled by a central help desk and are communicated to a single point of contact at the telephone company.

The largest CENTREX system for Wells Fargo services 5000 stations at 20 different locations in downtown San Francisco. This system was converted from an AT&T analog electronic central office to a Northern Telecom DMS-100 in September 1984 and was thus one of the first digital CENTREX services to be implemented in the US. The transition was completed and tested very smoothly over one weekend, as the bank kept its same rented telephone sets, 1A2 key telephone systems, and in-building wiring from the earlier CENTREX service. The set of features available to the users, including automatic route selection, and the total system costs remained essentially the same after this changeover at the central office. The telephone company and the bank each provided one person in every one of the 20 office locations to coordinate the cutover and test the service.

In February 1988, PacBell initiated the move of the Wells Fargo CENTREX service from one central office to another DMS-100, in a building to the south of San Francisco's financial district. Prior to that move the bank had been paying off-premises extension charges for about 2500 lines. Since the telco was forced to make this move because of capacity problems at the original DMS-100, these OPX charges (that might have applied to the other 2500 lines) were suspended, for effective saving of $26,000 monthly.

This second changeover was also scheduled for one weekend, but in the week following the cutover some one thousand telephones suffered from lack of ringing and lamp signaling current. Unfortunately, the only tests that had been conducted during the cutover weekend were of dialing out and these did not reveal the problems that 20% of the users would have with incoming calls and which persisted for four working days before the fault was cleared.

The Wells Fargo Bank is now testing Customer System Administration with its main CENTREX system from a workstation within its telecommunication office. SMDR data tapes are provided by PacBell for all of the CENTREX systems on a monthly basis, and these data are processed with bank-owned software on IBM mainframe computers. CENTREX services are not being used for data communication and the bank does not have any ISDN trial underway or planned, but is monitoring these developments closely.

As Wells Fargo approaches the fourth anniversary of its implementation of digital CENTREX, a complete evaluation of the viable alternatives for telephone service in San Francisco is being made. At mid-1988 the *average* cost for each line on this CENTREX system was about $45 monthly, including the actual CENTREX tariff, the rental of on-premises telephone apparatus, and shared network usage costs.

PacBell did not offer long-term tariffs when digital CENTREX was first introduced, so a rate stability contract would now probably offer significant savings. Additional long-term cost reductions may be achieved by competitive purchasing of telephone sets and by replacement of the outdated key systems. However, if it is not possible to obtain worthwhile reductions in the per line cost for CENTREX then the alternative of acquiring a network of bank-owned PBX systems would seem to be more cost-effective.

7.3 IMPLEMENTING CENTREX AT THE TORONTO-DOMINION BANK

The Toronto-Dominion Bank chose to go with CENTREX III on the basis of its satisfaction with the telephone company's level of service and support and because of the potential for significant cost savings, compared with an existing system that did not have the features available with modern technology. The bank did not wish to have its own staff involved in the detailed administration of its telephone system and also wanted to have the same system in large urban areas across the country. Prior to the start of the change to CENTREX III, in September 1985, Toronto Dominion had used CENTREX I (based on a crossbar central office) for its head office.

In Toronto the head office of the bank is now distributed throughout four office towers, which are on two adjacent city blocks and are shared with numerous other tenants, and several hundred people work at the Corporate Computing Facility at another building which is about one mile (1.5 km) away. CENTREX III service is also provided to 19 bank branches that are in the downtown area served by the same CENTREX host.

The cutover to CENTREX III was especially complex for Toronto Dominion because it coincided with the move of a large number of office personnel to a new building (known as the IBM Tower). A total of 3394 lines, that would ultimately support a total of 3460 telephone sets, was provided with dial tone by CENTREX III on September 3, 1985. In addition to DID to each line, the bank started with four major features, namely automatic route selection, ring again, call transfer, and three-party conference. The conferencing capability was subsequently extended to the large conferencing feature (an extra cost item) for up to 18 lines on one conversation. The telephones were actually changed from CENTREX I to CENTREX III sets at an average rate of 125 each week from September 1985 to March 1986. Whenever it was appropriate the change of a telephone set was coordinated with a move of offices.

The great advantage of a phased cutover, with a maximum of 400 new telephones going in over one weekend, was that it allowed suitable training to be given to all those concerned in less than two weeks before the change. It would have been impossible to provide timely and satisfactory training to 3600 people with a so-called flash cut. A total of 1900 person-days was taken up with the bank's Toronto cutover to CENTREX III, of which 1000 days were accounted for by an average of two hours training for each user. A team of 150 bank employees was trained as conversion leaders by Telecom Canada. In addition, 27 conversion administrators were designated within the bank, with one for each division or major department.

This major cutover necessitated changing the first three digits of all of the telephone numbers involved in Toronto. The bank devoted considerable attention to providing information to its many customers, through printed messages on monthly statements, leaflets stuffed in statement mailings, advertisements in newspapers, and notices in the branches. In spite of these efforts, the telephone company's changed number intercept service recorded 28,000 calls to the old numbers on the first day after cutover. This volume of redirected calls quickly fell to 11,000 by the third day, but the recorded changed number announcement was kept on the old numbers for up to one year, until the Bell Canada telephone directories were updated with the new numbers.

The CENTREX implementation team at the Toronto-Dominion Bank set up cost-minimization guidelines to decide the allocation of the electronic business sets to less than 30% of the total number of telephones. The EBS is needed for two-line access and for easy use of some features, and may incorporate a small visual display as an option. In fact, the Toronto system now has 2680 Unity phones and 780 EBS, for an actual proportion of only 22% of the more expensive sets. Of the EBS units, 314 have the digit display option.

CENTREX III has now also been implemented in regional offices and downtown branches in Montreal, Winnipeg, Calgary, and Vancouver, serving a total of nearly 2300 lines in those four cities. The bank has received a proposal from Bell Canada for the implementation of citywide CENTREX in Toronto. This would be a five-phase cutover, dependent on the availability of digital public exchanges in the metropolitan area.

7.4 SAVING ON COSTS

The Toronto-Dominion Bank has eight voice tie lines between Toronto and Montreal and some foreign exchange, or "out-of-area" lines, to New York City. These tie and FX lines are accessible directly from the CENTREX systems. However, the bank relies on the use of wide-area telephone service (WATS) for the bulk of its outgoing long distance calling. The WATS virtual network illustrated

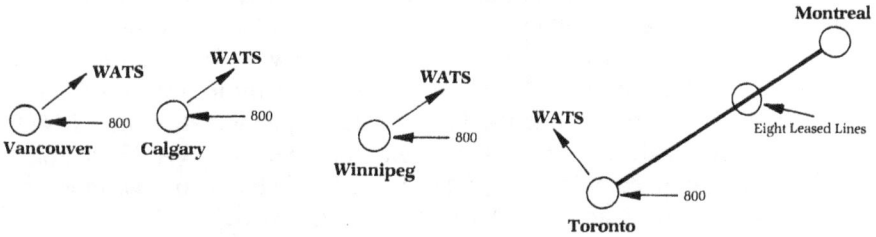

Figure 7.1 CENTREX locations and telephone network for Toronto-Dominion Bank.

in Figure 7.1 is carrying well over 30% more traffic than the volume originally estimated by the telephone company, in 1984, but has not been reconfigured since that time. WATS usage out of Toronto is now over 160,000 minutes per month, for a total monthly cost of about $76,000. Less than 1% of long-distance traffic overflows to full direct distance dialing (DDD) rates, as shown by recent in-Canada DDD billing of $1700 for one month in Toronto.

For a period of eight months, from June 1986 through to February 1987, a software problem in the CENTREX III system incorrectly marked some WATS (virtual) lines as being busy, due to improper use of the telephone set by users. This meant that calls were pushed into higher rates on other WATS lines. As is usual in these circumstances, the carrier initially blamed the user for this costly problem, but made full refunds when the bug was identified and fixed.

A major advantage with digital CENTREX is that all chargeable calls can be tracked, compared with the necessity of placing long-distance calls through the switchboard with CENTREX I if a billing record was needed. The IBM-compatible call detail tapes, produced by the DMS-100 exchanges, are sent by the telephone company to a service bureau that prints detailed reports under contract to the bank. A report of long-distance usage and extension details is sent to each departmental or branch manager. This report identifies any usage of DDD, as compared with WATS or leased lines, and flags the extra cost involved. Summary reports for cost centers and divisions are also produced.

The automatic route selection feature of CENTREX III is used extensively. Nine different classes of service (COS) have been established, ranging from local calling area service only to unrestricted DDD availability. The COS codes are decided at the departmental level at station review time and no central records of how COS are distributed across all the users are kept. Fewer than 100 persons in the bank, starting at the general manager or vice-president level, have access to DDD without the most-expensive-route warning tone.

The bank had six attendant console positions with the old CENTREX system and started with four consoles at the cutover to CENTREX III. The centralized

attendant service for all of its Toronto locations (i.e., over 3600 telephones) now
has only three consoles, staffed by three full-time and a few part-time attendants.
This efficiency can be attributed to all of the lines having DID and the 800-code
lines (previously known as INWATS) also going directly to groups of lines, by-
passing the switchboard attendants. Prior to the implementation of CENTREX
III a number of departments within the bank had their own WATS facilities. All
WATS usage is now combined through digital CENTREX to give much better
utilization of the system and so reduce long-distance costs.

7.5 OTHER OPTIONS

The Toronto-Dominion Bank has been consistent in managing its telephone
system carefully, while relying on Bell Canada whenever that is cost-effective. Bell
technicians do all of the additions, moves, and changes for the CENTREX system.
The proposed customer management system does not yet appear particularly attractive, because its capability would be limited initially to the single-line Unity
telephone sets.

There is no use of CENTREX for data communication purposes, even though
this capability was one of the lesser reasons for choosing CENTREX.

After some significant problems were overcome in the first 18 months of
digital CENTREX service, it is clear that Toronto-Dominion now has a telephone
system that meets all of the bank's needs and that the users are very satisfied with
the quality of service provided by the telephone company.

7.6 CITIBANK REJECTS CENTREX SERVICE

Citibank Canada, which is a subsidiary of New York-based Citicorp, has
nearly one thousand telephones in its new office building in downtown Toronto.
In November, 1985, the senior management of Citibank, with the concurrence of
its telecommunication consultant, decided in favor of CENTREX III to serve its
offices. The two factors that strongly influenced this decision were the lack of
personnel and expertise in telecommunication at that time and the prospect that
some business units of Citibank might be moving out of the downtown core within
two years.

The cutover to CENTREX was completed in a businesslike manner and went
very smoothly in April of 1986. This good experience was in contrast to the problems that were encountered by several large Canadian banks that switched to digital
CENTREX in the previous year. The bank has had no problems with the DMS-
100 host and finds that digital CENTREX is extremely reliable for voice communication.

However, Citibank has experienced serious problems with the system administration for CENTREX that is provided by Bell Canada. In the eighteen months up to February 1988, the bank's records show that 48% of the orders for moves, adds, and changes that were sent to Bell were done incorrectly. Even after these problems were known at the vice-presidential level within Bell Canada, there was still a delay of about six months before the system administration improved significantly.

In the spring of 1988 the bank continued to hold back part of the monthly payment for CENTREX in order to register its discontent with the service. At the same time the telecommunication department was in the process of writing a request for information to PBX vendors. This group believes that digital CENTREX is from 12 to 18 months behind the comparable PBX systems (e.g., the SL-1 or IBM 9750) in the delivery of features and applications. For example, a voice messaging system that is fully integrated with Meridian CENTREX was not yet available by mid-1988.

Citibank is locked into a three-year contract with Bell Canada for CENTREX, in the same way that other customers are, but is now actively planning to regain the management of its telecommunication network once this becomes feasible. The bank definitely wishes to control the moves, adds, and changes on its system and to be able to implement a fully integrated voice-data network. The concept of a CO-LAN, based on CENTREX does not really interest Citibank's management, as they are not willing to ship data out to and back from a central office, which is several city blocks to the north of their offices, as a substitute for an in-house network. The personnel requirement for the administration of a bank-owned PBX is certainly no more onerous than the task of supervising work that was being poorly executed by telco technicians. In both cases it is expected to require one administrator for the 1000 directory numbers.

Another consideration at Citibank, which is common to the many organizations with an emphasis on IBM-provided information-processing systems, is the desire to install IBM's Type 2 (shielded twisted pair) data wiring. In the Canadian context these plans for an integrated cabling system that can potentially support token-passing ring LANs at 16 Mb/s are incompatible with CENTREX service rented from Bell.

This situation clearly illustrates that the future prospects for CENTREX may be somewhat clouded, both by mediocre service from the carrier and the expectations that large, forward-looking organizations have for local area networks. The CENTREX *versus* PBX-LAN considerations need to be thoroughly reevaluated every year for most organizations.

Chapter 8
CENTREX in Retailing

8.1 RETAIL CHAINS

Chains of retail stores are one of the three main industry groupings that are the targets of the telephone companies for modern CENTREX service. These retail groups often have a number of stores in one urban area, together with warehouses and offices in other locations, combined with a very strict attitude to the nonproductive use of floor space. In addition there is usually a large volume of calls coming in from customers, for whom a common number and a centralized attendant service are valuable convenience and sales features.

8.2 NETWORKING ADVANTAGES

The T. Eaton company is the second-largest group of department stores in Canada, with major stores in many locations and warehouses in the major cities. Eaton's has followed a policy of installing Northern Telecom SL-1 PBX systems in its stores and offices outside the Toronto area. In Toronto the company has its head office, its flagship store (at the well-known Eaton Center), six other stores, and the national warehouse. The company started to reevaluate its telephone systems in the Toronto area in 1984, at which time the head office and main store were on CENTREX II service and the other locations used electromechanical Bell 701 PBXs. The warehouse-service center switchboard was then grossly overloaded and the in-store systems were becoming rundown and difficult to maintain.

A financial analysis of the alternatives for telephone service for Eaton's in Toronto showed clearly that it would not be advantageous to rent CENTREX on the basis of a stand-alone system for an individual building. In that situation a purchased PBX could show a pay-back against CENTREX in a little over two years. However, the citywide potential of CENTREX showed a different financial picture. Two other significant advantages for CENTREX III, in 1985, were the likelihood that many of the head office personnel would be moved within three years (from the office tower at the northern end of the Eaton Center) and that in the downtown locations there was no space available for a PBX. By using CEN-TREX, Eaton's could avoid the expensive problems involved in moving and splitting a large PBX system.

Eaton's is now paying $30.70 (Canadian) per line for CENTREX III for the first 160 lines on one host exchange and $24.90 (Canadian) monthly for subsequent lines. Network access across the city through CENTREX costs only $2 per extension, as part of these costs. The CENTREX III package provides a full range of features and uses push-button (Touch-Tone) telephones. Call detail recording and automatic route selection each add an (optional) fifty cents to these rates, per line.

Nine attendant's consoles are located in the main store location, of which eight are in regular use to handle the traffic for 3500 telephones. This centralized attendant service saves over $300,000 annually in attendants' salaries, compared with the previously installed separate switchboards, and yet provides a much better service in quickly connecting an outside caller to the appropriate department.

The locations served by digital CENTREX for Eaton's in the Toronto area are shown in the outline map in Figure 8.1, where the number of telephones in each building is given in parentheses. CENTREX III service to all of these buildings was implemented between late 1985 and October 1986, except for the store at Sherway Gardens, where Bell Canada was short of cable pairs into the mall. This store was cutover in February 1987. The store in Sherway Gardens is actually 23,000 ft (4.4 mi or 7.1 km) from the serving central office which meant that the electronic business sets could not be used at all. The telecommunication manager at Eaton's is not totally satisfied with the performance of electronic key telephones as substitutes.

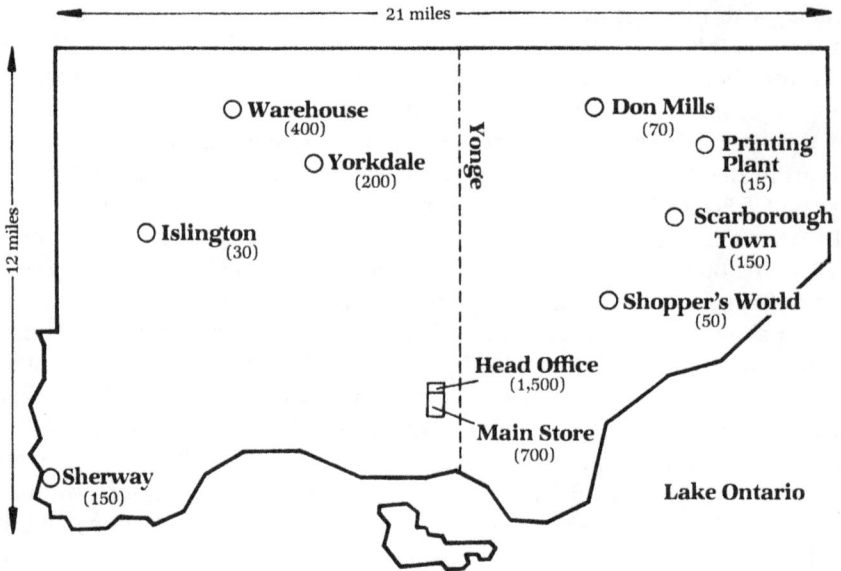

Figure 8.1 Eaton's locations served by CENTREX in Toronto.

8.3 LIMITATIONS AND PROBLEMS

Some serious problems were encountered with the new CENTREX service during 1986, but the software now appears to be stable. In the first year some EBS phones turned on at random, when they were not actually in use, and system partitioning within the DMS-100 between different users appeared to break down. This meant that an internal call could be going to an extension in the offices of a rival retailer.

There are some problems with record keeping inside the telephone company. Eaton's has encountered situations where a Bell technician has referred to an out-of-date equipment record, and then incorrectly changed, by a keyboard entry, the set of features associated with a number of telephones. Another problem is that the CENTREX switch cannot distinguish between individual trunks in a group used for 800 code service, so that load balancing and identification of the usage of various lines are difficult.

In general, record keeping and change management are easier with an in-house PBX than with CENTREX. Even when the user has the Customer Site Administration (CSA) subsystem for additions, moves, and changes, control will still be split with the telephone company. This means that a system administrator on the customer's site and a technician in the central office could be working at cross purposes.

The customer system management feature produces call detail tapes about ten days after the end of each month. These data are not the full complex record directly from the DMS-100 system, but are heavily preprocessed by the telephone company. The accounting cycle at Eaton's is based on four-week periods (not on a calendar month), which means that it is impossible to match the call detail reports from CENTREX with departmental and store accounting within the company. The real need is for on-demand call detail recording, with which the user can specify the "window" (i.e., the from and to dates) for the call detail to be recorded. As a management tool, Eaton's does not require full reports every month but rather wishes to more precisely spot check certain offices at intervals.

It was not possible to establish a rational citywide numbering scheme within Eaton's at the cutover to CENTREX III, for a number of reasons. The main store has 700 telephones, compared with 30 to 150 phones in the other Toronto-area stores. This means that a specific department in the downtown store may have ten telephones for the one set in the same department at a suburban store. It is clearly impossible to allocate blocks of numbers to similar functions in different locations under this condition. In addition there was a lot of resistance by long-term employees in the head office to a change of personal extension numbers. Various store managers were of the opinion that some customers had minor objections to the fact that the seven-digit number for a nearby store no longer had a local (recognizable) three-digit prefix, because all of the Eaton's numbers in Toronto started with the same prefix. There were no major problems in switching to a common numbering system with centralized console attendants, except for some misprints in the Toronto directory. The telecom manager at Eaton's hopes soon to be able to take advantage of the capability of the EBS to display the telephone number that is originating an incoming call or the name of a store on the alpha-numeric display panel on the telephone.

The billing minimum of 160 CENTREX lines on one host system (known as a "wire center") means that Eaton's is paying the full tariff cost for 160 telephones at the store in the Shopper's World Centre, Toronto, even though there are only 50 phones in that location. It would have been desirable also to connect a smaller store at the Rockwood Mall in Mississauga (immediately west of Toronto) into the CENTREX III network. This store is actually served by the same central office system that also serves the Sherway Gardens store. However, since Mississauga is a separate exchange area from Toronto, the switch is logically partitioned and the telephone company will not allow this connection.

8.4 EFFECT OF CENTREX

There were strong financial and convenience reasons for Eaton's to choose CENTREX III in Toronto and this has now proved to be the correct choice. Many of the head office personnel (with over 1000 phones) are expected to move from the downtown tower to the main warehouse site in northwest Toronto in 1989. Because of the CENTREX network the telecommunication aspects of this move will be fairly straightforward to manage. After the move has taken place it may well be cost-effective to use nailed-up (hot-line) connections through CENTREX for data communication purposes, as a lower-cost alternative to leased multichannel digital links (e.g., Megastream). The present monthly cost of $80 (Canadian) of a data interface module appears to be excessive, when this amount is added to the cost of a CENTREX line for data communication within the one building.

Experience with CENTREX III in Toronto has shown to Eaton's management the great advantage of using DID numbers for customer and employee-related services. As a result of this good experience the company has now put the DID feature on about 50% of the extensions on its PBX systems across the country.

Chapter 9
CENTREX in Government and Universities

9.1 MAJOR CENTREX MARKET

Governments, at the national, regional, and local levels, and government-financed institutions, such as universities and colleges, have always ranked as the top users of CENTREX services. The federal government of Canada has one of the largest CENTREX systems in North America, serving some 70,000 telephones in the Ottawa-Hull area. The older CENTREX services, with a very limited range of features, lost a considerable share of the market to digital PBX systems in the period from 1980 to 1985. This fall-off in CENTREX business was particularly significant in the university market, where there have been many sales of large integrated PBX systems in the last few years, especially for AT&T, InteCom, NEC, Northern Telecom, and IBM-Rolm. However, the four levels of government comprise, collectively, well over one-half of the total CENTREX market in North America.

9.2 STATEWIDE CENTREX

A good example of a government organization that has now made a firm commitment to modern CENTREX service is the state of Wisconsin. In 1986 the Department of Administration of the state, based in Madison, had to make a decision regarding the replacement of its old analog CENTREX services. A total of 35 CENTREX systems, based on crossbar or No. 1 ESS (electronic analog) exchanges provided CENTREX to 14,500 lines in state offices, 30,500 lines in universities, and also 15,500 lines in local agencies (such as county and city governments). All these organizations save money by having no local call charges for calls between agencies and by access to the State Telecommunication System (STS) network for reduced cost long-distance calling. Wisconsin Bell has now replaced its electromechanical and analog central offices with 5ESS and DMS-100 systems.

The state's CENTREX service is now provided on a 70:30 ratio by AT&T and Northern Telecom systems, with an identical set of features in all of the offices. The cutovers to digital CENTREX went somewhat more smoothly in those locations that use DMS-100 but most of the early problems with the statewide transition have now been cleared up.

The telecommunication department of the state of Wisconsin decided against issuing a formal request for proposal (RFP) to PBX vendors, on the basis of the time and costs required for creating the RFP and then evaluating the detailed proposals. The department estimated that following strict RFP bidding procedures for the customer premises equipment (i.e., PBXs and related apparatus) for the various locations would necessitate 12 person-years of work for state employees or consultants. This work would cost a total of over $1 million. The bidding process could also have taken an elapsed time of two years, meaning that the selected vendor could not have started to install new digital switches on government premises until June 1988.

9.3 CENTREX OR PBX

For comparison purposes the state's telecommunication analysts assumed an installed purchase price of about $540 for each telephone on a PBX and a monthly maintenance cost of less than $3 per line. The assumed per line cost is well below the present level of quotations for medium-to-large PBX systems, when telephone sets and cabling are included. A cost of $700 per line for a complete voice-only PBX is the generally accepted figure in the US.

Prior to 1986 the state of Wisconsin had been using a mixture of CENTREX services for 15 years. The state government is clearly both the largest customer and largest CENTREX user with Wisconsin Bell. The average cost of the existing CENTREX services was slightly over $14 per month for each telephone (at mid-1986).

The telephone company agreed to a rate stability contract at slightly less than $11 per month per line for the new digital CENTREX (which is known as CENTREX IV by Wisconsin Bell). This contract will run for seven years and will save the state some $28 million during that period, compared with the option of purchasing PBX systems at the lowest possible price. A further immediate saving was produced as Wisconsin Bell discounted the price of all existing CENTREX lines to a lower rate (from $14.95 to $12.65) as the conversion proceeded.

It could well be argued that the purchase price of a PBX is less than 40% of its total life-cycle cost, over seven years, when all other costs, such as system administration, maintenance, and network charges are taken into account. If this argument is accepted then the CENTREX IV option is still better for Wisconsin than receiving a gift of free PBX systems!

Up to 15,000 additional CENTREX IV lines will be installed in related government agencies at the under $11 per telephone price agreed upon in the state government's contract. There are 25 universities and community colleges in the state and several of these campuses have already installed digital CENTREX services.

The state has purchased all of the telephone sets that are connected to CEN-TREX IV. Most telephones are Comdial single-line (2500-type) sets, multiline feature phones are from Panasonic, and some old rotary dial (500-type) sets are still in use. The telecommunication adminstration is implementing a scheme for universal cabling, based on eight pairs of wire to two outlets on one face-plate for each user. Nearly 40% of CENTREX users, in four college campuses and one large office complex, are now connected to this standard wiring and cross-connect system, for both voice and data communication.

One major concern in the academic institutions is that the new CENTREX service has a maximum asynchronous data transmission speed of 19.2 kb/s. The large University of Wisconsin campus in Madison was close to issuing an RFP for an integrated voice-data-video switching system, when the decision to go with CENTREX IV was made by the state government in July 1986. It now seems likely that a separate data network may be essential for the university, to support the higher data rates that are needed for file transfers and high resolution display screens. Up to one quarter of the $4 million per year that will be saved from the conversion to CENTREX IV may be used to supply internal data communication systems to the larger universities in Wisconsin. Optical fiber systems will be installed throughout most campuses.

9.4 PERSONNEL CONSIDERATIONS

The Bureau of Information and Telecommunication Management for the state of Wisconsin is well aware that telephone systems do not manage themselves. Based on experience with a PBX that is installed at the Stevens Point campus of the University of Wisconsin, a ratio of one full-time employee (professional, technical, or clerical) per 1000 lines is a reasonable assumption. Because the local telephone company performs many of these functions with CENTREX, the personnel requirement for managing this large CENTREX system has been expected to be one full-time employee in the telecommunication department for every 2500 lines.

9.5 DIGITAL NETWORK

The enhanced CENTREX service was installed quickly by Wisconsin Bell, to the extent that 90% of the state government's telephone lines were connected to digital switches by the end of 1987. It is important to note that CENTREX IV provides a wide range of features to improve the productivity of office workers, and so upgrade service quality to the public, as well as offering significant cost savings.

In parallel with its mid-1986 decision in favor of CENTREX, the state also awarded a larger contract (for around $20 million each year) to AT&T for a digital network that will tie together 1800 government sites in Wisconsin. This integrated STS network is operated under contract by personnel from AT&T. The state government has no wish to operate its own telephone company. A network management center, with facilities identical to those used by AT&T for its own long-distance network, is based in Madison. The backbone of this network for 60,000 users consists of seventeen 1.544 Mb/s digital links (known as Accunet T1.5 by AT&T) and has a total of 2000 circuits. This new network will help the state avoid about $16 million per year in extra costs that resulted from recent changes in analog private line tariffs.

Trunks will be provided between the CENTREX systems throughout the state to the digital STS network in the range of one trunk for between seven and ten telephones, depending on the network traffic expected at each location. The ratio of trunks to lines on PBX systems is generally around 1:16. A large PBX was recently installed in a midwestern university with a trunk-to-line ratio of 1:13. The new CENTREX systems will be providing more generous long-distance network access than the typical PBX configuration that was used for comparison purposes in the financial evaluation.

CENTREX has proved to be the ideal vehicle to feed traffic into the long-distance network. At 23 sites the CENTREX node is effectively a concentrator for the STS. A payroll application from all of the state's campuses uses CENTREX to access a host computer in Madison. Transmission is at 9600 b/s through bit compression multiplexors.

The state of Wisconsin is typical of most large organizations in that it has numerous, separate data networks that crisscross like a multilayered spider's web. Four large departments had major dedicated star networks, of leased analog lines, based on Madison. A common digital data network is being installed by AT&T during the second half of 1988. The aims of this network are to achieve long-term cost containment and centralized performance monitoring. It seems likely that large-scale integration of the voice and data networks for the state will start in 1990.

9.6 LARGE-SCALE EFFECTIVENESS

Clearly, the state of Wisconsin, which accounts for 1% of the total annual revenues of Wisconsin Bell, was in a strong position to negotiate for an exceptionally good price on its CENTREX lines. It is more usual to expect CENTREX costs to be about $15 per line each month, rather than the close to $10 figure that applies to Wisconsin. At the same time, the effectiveness of modern CENTREX service combined with a digital network is obvious for an organization with many offices in cities and towns over an area as large as a midwestern state.

Now that most of the offices of the state of Wisconsin have been on digital CENTREX for a year, the state's telecommunication acquisition manager is daily convinced of the correctness of the choice as he sees the advantages of the service to its users and the effectiveness of having system administration under the telco's control.

9.7 CENTREX IN COUNTY GOVERNMENT

The local government of Iredell County, in North Carolina, provides a wide range of services to a population of nearly 90,000, with about 400 employees and some 30 different departments. Most of the county's office workers are in a number of buildings in Statesville and a few departments have branch offices in Mooresville, just to the north of Charlotte.

The main offices of Iredell County had an AT&T electromechanical PBX for about ten years, but suffered serious outage problems as the switchboard became older. The deputy finance officer for Iredell considered several types of digital PBX for the county, but soon realized that the cost of a network to link multiple PBX systems made digital CENTREX look very attractive. It was not necessary to go through a formal bidding process for CENTREX, known as ESSX by Southern Bell, as this is a service, not a system.

The cutover to ESSX was done in four stages, as contracts expired on older PBXs, over a total of 14 months up to March, 1988, for a total of 232 extensions. Southern Bell sent five instructors to provide classroom training and to work with heavy telephone users on a one-to-one basis. Each phase of the cutover went very smoothly, except for a problem with connections to foreign exchange lines to the offices in Mooresville, where service is provided by an independent telephone operating company.

ESSX is provided from a 5ESS central office that is situated only two blocks away from the busiest county offices, which are in three separate buildings. The cost for each ESSX line for these offices is $14.50 per month, with direct inward dialing, call forwarding, and excellent conferencing facilities for each telephone. Many of these lines become very busy at tax collection and election times and the DID feature then becomes particularly useful. Over 75% of incoming calls now bypass the attendant, so that the console position has become an information center. All of the telephone sets were purchased through a formal tendering process from a wholesaler and are maintained, at very competitive rates, by an independent service company. The Panasonic single-line sets cost $35 and the few speakerphone sets were about $50 each. A Tone-Commander console was purchased for

$900. Up to ten personal computers, each with an external modem, are connected to ESSX, together with two facsimile machines.

The county's Social Services department is housed in a building about one mile away from the serving central office and, with 85 telephones, has seriously outgrown the capacity of its AT&T Horizon telephone system. Unfortunately, this system was supplied on a long-term contract with a significant buyout penalty that is still outstanding and, in addition, DID service is not so critical to that department as to some others (such as tax collection). Under these circumstances it is impossible to recommend ESSX at $17.50 for each line and it seems that the social services building will have to be served by a new, small AT&T PBX before the end of 1988.

This example of Iredell County demonstrates that for a fairly small organization, scattered over a number of buildings in one town, CENTREX is generally cost-effective for 90% of the users, with the added advantages of unlimited growth capacity, complete DID service, and minimal requirements for administration, operation, and equipment accommodation. However, the tariff is still distance dependent and this means that where a large group, of 50 or more people, is more than a mile away from the central office and when a high proportion of those users cannot justify DID, then a digital PBX may still prove more economical.

Chapter 10
Implementing a CENTREX System

10.1 USERS' GROUPS

The experience of CENTREX users in Canada has shown that it is extremely useful to have a strong users' group to put pressure on the telephone companies when CENTREX service fails to meet its full potential. The users' group is composed of telecom managers representing the CENTREX customers and is closely affiliated with the association of business telephone users. The major objective of the CENTREX users' group is to resolve difficulties encountered with the service and to interact with the telephone companies and with manufacturers of CENTREX systems. The major Canadian telephone companies and the system supplier (Northern Telecom in this case) have designated specific liaison managers to work with the users' group. The Canadian CENTREX users' group currently has a membership of 26 companies, together with government agencies at the federal, provincial, and municipal levels. Several CENTREX users' groups are active in the US at the national and regional levels.

The users' group has an invaluable role in showing that serious problems are common to a number of users and are not unique to one customer. This activity counteracts the tendency of management at the telephone company to divide and conquer the customers. At the same time the telephone companies are finding that the users' group is useful as a forum in which to present ideas on future developments and features. This activity is largely done at scheduled meetings between the carrier, the system manufacturer, and the members of the users' group.

Two years ago the Canadian users' group was very disturbed with major shortcomings in the CENTREX III software and some of the apparatus, particularly the electronic business sets. The group is partly responsible for action being taken to rectify numerous problems. The list of problems and desirable enhancements is now shorter and less urgent, but is still significant.

The CENTREX users' group is still concerned about the quality of system administration for CENTREX in some areas of the country. The technicians working on the system appear to lack appropriate knowledge, thus causing software problems and being slow to rectify faults. Some of the major features promised for CENTREX III have not yet been delivered. For example, the Customer System Administration capability had been in field trials with one major user for 18 months and did not appear to have a firm due date for release at the end of 1987.

In general it appears that an *average* of 18 months elapses between the identification of a specific software problem and the solution to that problem being released by the telephone company. A *median* resolution time of nine months for significant problems seems to be a generally accepted figure.

10.2 TRAINING FOR CENTREX

All of the telecom managers who are involved with CENTREX systems emphasize that digital CENTREX can provide many of the same features as a sophisticated PBX and, therefore, fully justifies taking the same care with implementation and the training of users. A well thought out and thoroughly documented implementation plan is essential. If at all possible, a phased implementation schedule is preferable, to ease problems with training and assisting users immediately after system cutover.

An average of two hours of training on the telephone sets and features is needed for the users, a few days before cutover. Training should be done with small groups of people who will be using identical telephones and the same set of features. Larger organizations may find it desirable to create and use training material that is specific to their own applications. In CENTREX implementations it is fairly common for a small implementation team to train a large number of trainers within the organization, who in turn train the personnel in their specific department or location. This two-level training approach seems to be the most cost-effective solution to the provision of good quality training with a local flavor.

Adequate provision must also be made to train new personnel who move into the CENTREX-using offices and to refresh and enhance the training of all users as new features are added into the system.

10.3 INSTALLING A CENTREX SYSTEM

Numerous organizations have now gone through the process of installing large digital CENTREX systems, with thousands of users in some cases. Their cutover experiences can help with the planning and management of future CENTREX implementations.

Generally speaking a period of from a minimum of six months to a desirable 12 months should be allowed from between the time of decision to the actual switchover to digital CENTREX, whether from an older CENTREX service or from a PBX-based system. The implementation may include a change of central office code and the adoption of least-cost long-distance routing with, possibly, a new wide-area network. Complete utilization of direct inward dialing is most likely with CENTREX, and a new long-distance numbering plan may also be implemented.

Detailed station reviews with all future users of CENTREX are an essential aspect of the conversion. In this way the needs of each person can be determined for specific features, type of telephone, and call forwarding arrangements and this information is then used in programming the system. A three-person station review team, consisting of a telco representative, a telecommunication analyst, and an administrator from the specific user department, has been found to ensure the best results. Throughout the station review process the emphasis must be to provide to each user, only those features of CENTREX that have clearly identified benefits. The basic criterion in a station review is needs, not wants, and certainly not just a carrying forward of telephone type and features that are presently in use.

A freeze on any changes in the corporate telecommunication network for three months prior to a major CENTREX implementation is also very helpful. The station reviews should therefore be completed in a period of 60 days, leaving enough time for programming the system's data base. The detailed planning of a CENTREX implementation will probably require at least one full-time telecommunication expert for six months for every 1000 lines to be implemented.

As with the implementation of any new telephone system it is also essential to have accurate and detailed floor plans for all locations into which CENTREX service is going. Because a few moves and changes will have to be made during the 90-day freeze period, these must be carefully controlled and fully documented onto all of the plans.

The ratio of single-line telephone sets to more expensive multiline sets (perhaps with digital displays) in a digital CENTREX installation has a critical effect on the monthly cost of the system. The more expensive sets should go to those who need intercom service, to personnel who answer multiple lines, and to those who have exceptionally high call volumes. In many organizations the ratio of single-line to multiline sets on CENTREX will probably be around three to one. In small businesses with heavy customer contact the proportion of more expensive telephones may be justifiably higher.

As the date of a CENTREX cutover approaches, the inside users and the outside callers need to be educated about the coming changes. If these include a change of telephone number, perhaps as part of a move to a centralized, citywide system then this change demands maximum publicity. Experience has shown that many customers will initially attempt to use the old number, even when several publicity media have been used, such as newspaper advertisements, mailed notices, and paid announcements on local radio. This means that plans must be made with the telephone company to set up a generously staffed intercept service on the replaced numbers. The rate of calls on the old numbers falls off rapidly, within a few days following cutover, but does not drop completely to zero in less than a year from the date of change.

Training for all the users of a new digital CENTREX must be provided as close as possible in time prior to the conversion date. Training in small groups, of not more than 12 people in each group, should be done within ten working days of the change for the people in each group. Not only should all future CENTREX users be trained in how to use the specific telephone set which will be used but, in addition, digital CENTREX must be emphasized as a network service, that will help to integrate the organization and improve everyone's effectiveness.

During the actual switchover to digital CENTREX service the telco and the customer need a team of competent people, working in close cooperation, to handle the problems that inevitably happen with the start-up of any major, new system. Some of the early problems at cutover will be due to the lack of experience of the users with new telephones and additional features, while other difficulties may well be caused by a new software release or by new apparatus.

After the cutover of a new CENTREX system, rigorous post-implementation audits of system performance should be carried out. These audits test a significant sample of lines for dial-tone delay (DTD) in the busy hour, for traffic volumes on voice and data lines, and for activity levels at the attendant's console(s). As a good rule, busy hour DTD should not exceed three seconds on more than 1% of the locals and attendant's time utilization should not be more than 80% of available time, again in the busy hour of the week. The system audit should also review the utilization of station and system features by the users, to make sure that the availability and use are matched to needs.

The first CENTREX system audit could be one month after cutover, followed by a second audit at the three-month point and regular audits at six- to 12-month intervals from then onward. In this way the system can be kept most effective and data can be collected to help plan the next generation system.

Appendix A
Electronic Key Telephone Systems for CENTREX

This is *not* a complete list of EKTS suppliers. The systems from these companies have been used in most installations where EKTs have been combined with CENTREX.

AT&T Info Systems (Merlin/Spirit)
1 Speedwell Avenue   (201) 898-2000
Morristown, NJ 07960

AT&T Canada Inc.
3650 Victoria Park Ave., Suite 800
Willowdale, Ont.   (416) 499-9400
M2H 3P7

Contel IPC (Center Max)
600 Steamboat Road   (203) 661-7500
Greenwich, CT 06830

Eagle Telephonics Inc. (Eagle/One)
375 Orser Avenue   (516) 273-6700
Hauppage, NY 11788

Iwatsu America Inc. (Omega)
430 Commerce St.   (201) 935-8580
Carlstadt, NJ 07032

Iwatsu America Inc.
P.O. Box 2226   (416) 897-1755
Square One
Mississauga, Ont.
L5B 3L7

Northern Telecom Inc. (Vantage/Norstar)
565 Mariott Drive   (615) 885-3510
Nashville, TN 37210   or (800) 321-2649

Northern Telecom Canada Limited
2920 Matheson Boulevard East   (416) 328-7000
Mississauga, Ont.
L4W 4M7

Plant Equipment Inc.
28075 Diaz Road   (714) 676-4802
Tenecula, CA 92390

Tadiran Electronic Industries Inc.
Telecom Division
5733 Myerlake Circle   (813) 536-3220
Rubin Icot Center
Clearwater, FL 33520

TIE/Communications Inc. (ONYX)   (203) 926-2000
8 Progress Drive
Shelton, CT 06484

TIE/Communications Canada Inc.   (416) 475-5577
7550 Birchmount Road
Markham, Ont.
L3R 6C6

Trillium Telephone Systems Inc.
1675 MacArthur Boulevard   (714) 557-3300
Costa Mesa, CA 92626

Trillium Telephone Systems Ltd.
P.O. Box 13030
603 March Road   (613) 592-2550
Kanata, Ont.
K2K 1X3

Walker Telecommunications Corp. (Marathon CTX)
200 Orser Avenue   (516) 435-1100
Hauppage, NY 11788

Appendix B

Management Systems and Packages: Suppliers

American Telecorp (Cenpac)
10 Twin Dolphin Drive   (415) 595-7000
Redwood City, CA 94065

AT&T Network Systems
1 Speedwell Road   (201) 898-2000
Morristown, NJ 07960

AT&T Canada Inc.
3650 Victoria Park Ave., Suite 800
Willowdale, Ont.   (416) 449-9400
M2H 3P7

Conveyant Systems Inc. (Teledesk)   (714) 660-1801
2852 Alton Avenue
Irvine, CA 92714

GST-Telematic Inc. (Centran)
19204 North Creek Parkway   (206) 487-4333
Bothwell, WA 98011

GTE Communication Systems Inc.
2500 West Utopia Road   (602) 581-4351
Phoenix, AZ 85027

Microtel Ltd. (GTE)
100 Strowger Boulevard   (613) 342-6621
Belleville, Ont.
K6V 5W8

Moscom Corporation (M 3000)
300 Main Street   (716) 385-6440
East Rochester, NY 14445

NEC America (Astra-Phacs)
8 Old Sod Farm Road   (516) 473-7570
Melville, NY 11747

NEC Canada Inc.
6711 Mississauga Road   (416) 858-3500
Mississauga, Ont.
L5N 2W3

Northern Telecom Inc. (NM-1)
P.O. Box 13010   (919) 549-5000
4001 E. Chapel Hill — Nelson Hwy.
Research Triangle Park, NC 27709

Northern Telecom Canada Limited
2920 Matheson Boulevard East   (416) 238-7000
Mississauga, Ont.
L4W 4M7

Stromberg-Carlson Inc.
400 Rinehart Road   (305) 849-3000
Lake Mary, FL 32746

Siemens Information Systems Inc.
5500 Broken Sound   (305) 994-7232
Boca Raton, FL 33431

Telco Research Corp. (TRU)   (615) 329-0031
1207 17th Avenue South
Nashville, TN 37212

Telco Research Canada Ltd.
Madison Centre, #1004   (416) 733-0181
4950 Yonge Street
Toronto, Ont.
M2N 6K1

Appendix C
Voice Messaging Systems for CENTREX

AT&T Information Systems (Audix)
1 Speedwell Avenue   (201) 898-2000
Morristown, NJ 07960

Centigram Corporation (Voice Memo II)
4415 Fortran Court   (800) 942-4942
San Jose, CA 95134

Digital Sound Corporation
2030 Alameda Padre Sierra   (805) 569-0700
Santa Barbara, CA 93103

IBM-Rolm Division (Phonemail)
4900 Old Ironsides Boulevard   (408) 986-1000
Santa Clara, CA 95050

IBM Canada — Rolm Division
4 Lansing Square   (416) 296-6672
Willowdale, Ont.
M2J 1T1

Northern Telecom Inc. (Meridian Mail)
2305 Mission College   (408) 988-5550
Santa Clara, CA 95054

Northern Telecom Canada Limited
2920 Matheson Boulevard East   (416) 238-7000
Mississauga, Ont.
L4W 4M7

Octel Communications Corporation (Aspen)
1841 Zanker Road   (408) 947-4500
San Jose, CA 95112

Opcom Voice Messaging Systems (DIAL)
110 Rose Orchard Way   (800) 553-3425
San Jose, CA 95134

Call Pro
7100 Woodbine Avenue, Suite 106   (416) 470-7100
Markham, Ont.
L3R 5J2

VMX Inc. (IVMS)
1241 Columbia Drive   (214) 699-1461
Richardson, TX 75081

Appendix D
Local Transmission and Remote Systems for CENTREX

Alcatel Network Systems Inc. (1218 DLC)
3128 Smoketree Court   (919) 850-6000
Raleigh, NC 27604

AT&T Network Systems (SLC-96)
1 Speedwell Road   (201) 898-2000
Morristown, NJ 07960

NEC America Inc. (NEC 192/135 FOTS digital loop)
Transmission Systems Group   (703) 698-5540
2740 Prosperity Avenue
Fairfax, VA 22031

Northern Telecom Inc. (FMT 150)
P.O. Box 13010   (919) 549-5000
4001 E. Chapel Hill — Nelson Hwy.
Research Triangle Park, NC 27709

Northern Telecom Canada Limited
2920 Matheson Boulevard East   (416) 238-7000
Mississauga, Ont.
L4W 4M7

Appendix E
Glossary of CENTREX System Features

Automatic Call Distribution (ACD) is a system feature that provides the even routing of incoming calls to a group of telephones.

Automatic Route Selection (ARS) is a system feature that allows an end user to preselect the alternate facilities over which a long-distance call can be routed.

Basic Queuing allows a caller to line up for a busy long-distance facility. When a line in the long-distance facility becomes free, the caller is signaled and the call is automatically processed.

Call Back Queuing allows a caller to respond to a busy signal by invoking the queuing feature, hanging up, and then waiting for a signal that the call is ready to be processed. Once the busy line becomes free and the caller responds, then the call is automatically processed. It allows normal use of the telephone during the waiting period.

Call Forwarding allows a user to forward incoming calls to another telephone. There are several types, such as call forwarding–busy and call forwarding–no answer.

Call Park is a station feature that allows a user to hold a call coming into the system or one originated within the system. The call can be retrieved from any telephone on the system by dialing specific codes.

Call Pickup allows a user to answer an incoming call to another extension from the user's telephone, by dialing a code or using a call pickup button, if available.

Call Request allows an internal caller, upon receiving a busy or no answer, to dial an activation code, which lights the called party's "message waiting" lamp. When the called person later dials the call request retrieval code this automatically makes a connection to the original calling party's extension.

Call Transfer is a system feature that allows a user to transfer a call that is being established to another telephone number.

Call Waiting allows a user to be notified of a second incoming call, while on the line. This allows the second call to be answered while the first call is held, without the need for a second line on the telephone.

Centralized Attendant Service enables a single console location to answer and direct calls for a number of different locations in a city, or across a wider area.

Class of Service (COS) is the set of parameters that is defined for a user's extension that determines which lines and features that telephone may use. COS can be customized to each user and so greatly improves security and cost control.

Conferencing allows conference calls between groups of users and outside lines, for a maximum of 30 conferees.

Coordinated Dialing Plan allows any user in a system to call an extension at any other location within a network by dialing the same number.

Digital Recorded Announcement provides a recorded voice announcement, from a digital memory module, which is activated under certain calling conditions (e.g., when all attendants are busy).

Direct Inward Dialing (DID) is a system feature to allow incoming calls to be dialed directly to the user's extension.

Direct Inward System Access (DISA) allows an outside caller to gain access to network features and services as if the user were dialing from inside the system. DISA is carefully controlled by access codes and passwords.

Group Intercom is a system feature that allows abbreviated dialing among a specified group of telephones on the system.

Hunting allows calls to be redirected under busy conditions to alternate stations or available trunks within a specified group.

Message Center allows specified groups of lines to be routed to a specially configured telephone set, on a busy or no answer condition. The message center telephone is equipped to expedite the processing of calls, taking of messages, and activating or deactivating of message lights on individual stations.

Multiple Appearance Directory Number (MADN) is a system feature that allows a single extension number to be assigned to more than one physical telephone. MADNs can be arranged to ring busy when any of the appearances is in use or to allow incoming and outgoing calls from each telephone regardless of the status of the other phones.

Ring Again is a system feature that allows a caller to enter a queue for a busy number within the system. When the line becomes free, the caller is signaled and upon the caller's response the call is automatically processed.

Speed Calling allows users to dial short one- to four-digit codes to reach preprogrammed numbers. This feature is more accurately described as abbreviated dialing.

Uniform Call Distribution (UCD) allows incoming calls to be evenly distributed between a group of extensions that activate the UCD feature. Each station that is part of a UCD group may be reached in one of two ways: by dialing the station number; or by dialing the UCD number upon which the feature rings the station that has been on-hook for the longest time. The UCD feature is useful for allowing single-line telephones to operate like multiline sets and for incoming call load sharing at peak traffic times.

Note: Most of the definitions in Appendix E were adapted from the Meridian Services, *Reference Guide and Glossary,* which is distributed by Northern Telecom Canada.

List of Acronyms & Abbreviations

Acronym or Abbreviation	Definition
A channel	analog channel
ABS	Advanced Business Services (GTE)
ACD	automatic call distribution
AILC	asynchronous interface line card
AM	administration module
AP	applications processor
ARS	automatic route selection
AXE	automatic exchange (Ericsson)
B channel	bearer channel
BCS	batch change supplements
BHCA	busy hour call attempts
BMC	Billing Media Controller (Northern Telecom)
BOC	Bell operating company
BRCS	Business & Residence Custom Services
BU	base unit
CCITT	Consultative Committee on International Telephone and Telegraph
CCS	common channel signaling
ccs	hundred call seconds per hour
CCSA	Common Control Switching Arrangement (AT&T)
CDR	call detail recording
CENTREX	central exchange
CF	call forwarding
CM	communication module
CO	central office
CO-LAN	Central-office based LAN
COS	classes of service
CPE	customer's premises equipment
CSA	Customer Site Administration (Northern Telecom)
CSR	customer station rearrangement
CUI	control unit interface

Acronym or Abbreviation	Definition
D channel	data channel
DCO	Digital Central Office (Stromberg-Carlson)
DCO-SE	DCO-small equipment (Stromberg-Carlson)
DDD	direct distance dialing
DID	direct inward dialing
DISA	direct inward system access
DLC	digital line card
DMS	digital multiplexed system
DMS-100	Digital Multiplexed System-100 (Northern Telecom)
DNC	Dynamic Network Controller (Meridian)
DNX	Digital Network Exchange (Meridian)
DOD	direct outward dialing
DOV	data over voice
DP	data processing
DTD	dial-tone delay
DTMF	dual-tone multifrequency ("touch-tone" telephone)
EAX	Electronic Automatic Exchange
EBS	Electronic Business Set (Meridian)
EDP	electronic data processing
EKTS	electronic key telephone systems
ESN	see MSN
ESP	Extended Service Package (Stromberg-Carlson)
ESS	Electronic Switching System
ET	exchange termination
ETN	electronic tandem network
EWSD	electronic digital system (Siemens)
5EAX	No. 5 EAX (GTE)
5ESS	No. 5 ESS (AT&T)
FX	foreign exchange
GEAC	manufacturer of library automation systems (Canada)
H channel	high-speed channel
HICOM	digital PBX (Siemens)
IDN	integrated digital network
ISDN	integrated services digital network
ISLU	Integrated Services Line Unit (AT&T)
ISN	Information Systems Network (AT&T)
LAN	local area network
LAP-D	link access protocol type D
LC	line card
LPP	link peripheral processors
LT	loop termination

Acronym or Abbreviation	Definition
MADN	multiple appearance directory number
MAN	metropolitan area network
MIS	management information systems
MSN	Meridian Switched Network (formerly ESN)
NEAX 61E	digital central office system (NEC)
NEC	Nippon Electronic Company (Japan)
NT	network termination
OAM	operations, administration, & maintenance
OMNI(PBX)	digital PBX system (Fujitsu-GTE)
OPX	off-premises extensions
ORM	optical remote module
OSI	open systems interconnection
PABX	private automated branch exchange (PABX = PBX)
PBX	private branch exchange (PBX = PABX)
PDS	Premises Distribution System (AT&T)
POTS	plain old telephone service
RFI	request for information
RFP	request for proposal
RLCM	remote line concentrating module
RLU	remote line unit
RSLE	remote subscriber line equipment
RSLM	remote subscriber line module
RSM	remote switching module
RSU	remote switching unit
SDN	software defined network
SM	switching modules
SLC	Subscriber Line Carrier (AT&T)
SMDR	station message detail recording
ST	subscriber terminal
STS	State Telecommunication System
TA	terminal adapter
TCM	time compression modulation
TCMS	Telephone Cost Management System (Moscom)
TE	terminal equipment
telco	telephone operating company
TIF	terminal interface
TMS	time multiplexed switch
TRU	call detail recording system (Telco)
UCD	uniform call distribution
UNMA	Unified Network Management Architecture (AT&T)

Acronym or Abbreviation	Definition
USNBO	US National Bank of Oregon
WAN	wide-area network
WATS	wide-area telephone service

www.ingramcontent.com/pod-product-compliance
Lightning Source LLC
Chambersburg PA
CBHW021433180326
41458CB00001B/247